DISCARDED
Fordham University Libraries

MIND-BODY: A CATEGORIAL RELATION

MIND-BODY:
A CATEGORIAL RELATION

by

H. TRISTRAM ENGELHARDT, Jr.

MARTINUS NIJHOFF / THE HAGUE / 1973

© 1973 by Martinus Nijhoff, The Hague, Netherlands
All rights reserved, including the right to translate or to
reproduce this book or parts thereof in any form

ISBN 90 247 1550 4

PRINTED IN THE NETHERLANDS

CONTENTS

Preface	vii
I. Introduction	1
A. Occasions for an Investigation	1
B. Categories and Categorial Accounts	6
C. Programs of Investigation	13
D. Legitimacy of This Investigation	27
II. A Phenomenology of Mind and Body	28
A. Experience of Mind-Body	29
B. A Phenomenological Outline of an Ontology	60
III. Alternative Accounts	63
A. Conflicting Ontologies	63
B. Transcendental Requirements	85
IV. A Transcendental Ontological Account	89
A. A Dialectical Relation	89
B. The Dialectic of Mind and Body	105
C. Negative and Positive Dialectics and the Identity in Difference	119
D. An Answer to the *Quid Juris*	123
V. Ontological and Empirical Structures	126
A. Transcendental and Empirical Science	128
B. The Mind's Embodiment	130

    C. Structural Integration and Independence of Mind and Body   139
    D. Psyche and Soma   148
    E. Conclusion   161

INDEX   169

PREFACE

The relation of mind and body is one of the central problems of post-Cartesian times. It has precluded a unified theory of the positive sciences and prevented a satisfactory notion of man's psychophysical unity. Generally it has been treated as a problem of causality and solutions have been sought in various schemata of etiological relations. Proposals have ranged from that of reciprocal action between two substances and two causal streams to a reduction of all phenomena to a single causal stream involving a single class of substances. This investigation will abandon such schemata and attempt to start afresh. It will analyze the relation of strata of meaning involved and will be only tangentially concerned with the causal relations of mind and body.

This investigation will view the relation of mind and body no longer as the association of two substances, two things, but as the integration of two levels of conceptual richness. This is a move from hypostatization, reification, to categorialization – a move from the opacity of things to the relative lucidity of their significance. It recognizes that philosophy seeks not new facts about being but rather a way of understanding the integration of widely diverse domains of facts. Here the goal is the explanation of the unity of being, specifically the being of mind and body, in terms of thought – that for which being has significance and that for which incongruities of significance appear as a problem. The issue then becomes that of analyzing the conditions for the unity of two categories in thought, not that of determining the mechanism for the causal integration of two things. The point of departure chosen is phenomenological. The logic rooted in the matter of being is not empty and formal; it concerns the structure of reality. Phenomenology offers itself as a means for discovering categories – concepts essentially ingredient in appearance and apprehendable in thought. Or put another way, this investigation starts with ordinary experience, which provides a ready and accessible point of departure for the analysis of the presented conceptual structures of existence, the fabric of meanings in appearance which reason can recognize as its own. In short, this work is an endeavor to see questions concerning the structure of being as questions concerning the structure of concrete thought and thus to operate in a domain truly congenial to thought. Thought relieved of the burden of divining the comportment of things can concern itself with the logic of categories.

This approach shows the influence of philosophers such as Hegel, Husserl, and Kant. Though the phenomenological starting point is reminiscent of Husserl if not also of Hegel, it must be understood in its own setting. Similarly, though many concepts have been borrowed from Hegel, they have been altered and employed in ways that are often unorthodox. My in some respects Kantian use of "transcendental" should be an obvious indication of at least a terminological departure from Hegel. But neither is this investigation Kantian. Kantian terminology (e.g., "transcendental," "necessary conditions for the possibility of") appears in a transformed state. Unless otherwise noted, my terminology finds its sense solely in the context of this investigation. In short, this work must stand by itself. Hegel, Husserl, and Kant have been drawn upon only insofar as they provided ideas and suggestions that could be adapted to my needs. In this process of adaptation much was added and everything to some extent changed. The Hegelian project of a pure categorial ontology has been combined with an Husserlian project of discovering concepts through a phenomenology. This combination has then been used to outline the necessary limits and conditions of physiological and psychological accounts of man. Finally, despite the long history of the mind-body debate, the approach will not be historical. This is reflected in the policy of citation. References will be given only when they afford clearly pertinent elaborations of a point, or when the choice of terms or procedures was directly influenced by a particular work. Though I am indebted to many philosophers, I have developed and blended their insights so that most references would require a commentary in order not to imply or suggest a misreading of their works. In particular, I trust that a declaration of my deep indebtedness to Hegel, Husserl, and Kant will alert the reader to their influences.

What success this investigation may have is due to the guidance and criticism of Professors Irwin C. Lieb, Klaus Hartmann, Charles Hartshorne, John N. Findlay, and Richard M. Zaner. They contributed generously in time and insight to the doctoral dissertation, which was the ancestor of this work. But where this work may falter, this cannot be imputed to them. I am very grateful for a Fulbright scholarship (1969-70), which allowed the development of the dissertation into the present work, as well as for the close and beneficial association with Professor Klaus Hartmann during that period.

GALVESTON, TEXAS
July 29, 1972

CHAPTER I

INTRODUCTION

A. OCCASIONS FOR AN INVESTIGATION

1. The investigation's origin in ordinary experience

A philosopher begins the investigation of the relation of mind and body with a question already adumbrated by everyday experience. Man finds himself in a world that both aids and hinders his thoughts and actions. His thoughts and actions are on one side, set over against the forces and inertias of the world in which they must be actualized. One has plans which can be executed only to the extent that the objective world allows this. One has thoughts which only with effort and the proper concatenation of circumstances can, by means of the material, energies, and laws of the objective world, be turned into things. Between oneself and this objective world there stands one's living body – the crucial juncture between ideas and the physical reality in which they are realized.

Here within this objective world and through one's body, one meets his fellow men. The other whom one loves, hates, works with, and plans with is presented through his body – never immediately with thought touching naked thought, volition touching naked volition. Being a man among other men reveals that men are not a society of unembodied spirits. Nor am I such a spirit. Thoughts, volitions, and desires are actualized always in and through the body. The body, in short, reveals itself as one's embodiment in the world of physical objects as well as in the society of other men.

One's embodiment presents itself in and through one's seeing, hearing, tasting, smelling, touching, heat-sensing, position-sensing, hunger-sensing, thirsting, etc. All of these represent an involvement with objects and/or conditions in the world. And these involvements express a unity of mind and body. When one is thirsty, one asserts "I am thirsty," not "my body is thirsty." At least the naive position is that *one* feels, sees, hears, reproduces, ingests, not that one's body sees, hears, etc. And the same for others; to know a person is to know him, not just his body, nor usually just his mind without his body.

Yet there is a dichotomy, for thoughts and volitions are often set in opposition to the mute, cognitional and volitional opaqueness of the body. I want to move my foot but it is asleep; the blind man endeavors to see but is frustrated; the mortally ill endeavor to live, but die; when intoxicated one endeavors to think, but somehow the thoughts slip by. Ordinary experience offers contrasts engendered by the body's recalcitrance before one's thoughts and volitions. In addition, physical forces present the body as another object in the world, opposed or indifferent to thought and volition. The wind pushes me down the street when it pushes my body; the heat makes me perspire and be thirsty. And one finds this same dichotomy in others: one can desire a person's body without regard to his mind. One can admire a person's mind while being revolted by his bodily condition. Moreover, one can, when observing other persons in this world, observe the attempts of their minds to subdue the forces of this world – at times with heroic, at times with tragic consequences.

Ordinary experience thus presents the embodiment and the unity of man, as well as the dichotomy between the physical and psychical. Mind and body show themselves on inspection to be correlative and contrasting. With mind, one encounters a cognitional translucency opposed to the opacity and recalcitrance of the body. The experiences of sickness and physical limitation present the body as an other which the mind is always endeavoring to make its own. In fine, the mind and body contrast: the body is an other to the mind insofar as it is physical; the mental is other than the body insofar as the mind remains ideal and thus not externalized. The question then is, in what respect is one's body his and in what respect is it an object among others and opposable to the mind? The question concerns the nature of the relation of mind and body.

2. *Experience and physiology*

Physiology adds, as it were, a further dimension. The previously vague notions of dependencies become clarified in a manner that is startling in compass and clarity. We can know a friend paralyzed on one side because of an occlusion of the middle cerebral artery; we can know it, we can see it in angiograms as well as note the destruction of tissue by the lack of activity recorded in electroencephalograms. Or perhaps we encounter the entire disruption of his personality; we can see it is due to a tumor compressing the frontal lobes. We know the effects of drugs we take. We can say, "Now I'm sleepy. That is due to the fact that for many reasons my reticular activating formation is inhibited. As a result I concentrate on the paper and my writing only with difficulty and most imperfectly. What I'll do is drink some coffee – the caffeine will stimulate my reticular formation and this will excite the higher centers of my cerebral cortex (in coordination with the thalamus, etc.). I know how this works in cat brains because I have watched them become better and more discriminatory responders to stimuli when I have applied stimulation to the reticular activating formation. Moreover, I know many of the respects in which cat brains and human brains are alike; besides, I know of similar experiments in humans." Having drunk the coffee and become more alert, I happily conclude that I have applied sufficient stimulation. The result is that one can turn to one's mind as one would to a radio and proceed to tune it.[1] One can even perform on oneself the neurological experiments that Wittgenstein outlines in the *Blue Book*;[2] one can stimulate or inhibit portions of one's own brain and live through the "effects."

Now what does all this signify? First of all, it shows that there are surroundings, circumstances, in which one can and does talk about the mind being dependent upon the body. There is a proper use of such expressions when one, for example, manipulates his state of consciousness through coffee, a martini, etc., calling to mind the neurophysiological or, better, neuropharmacological processes and events involved in such endeavors. In short, there is a whole family of uses of words and expressions associated with instances of psychophysical dependency. These uses are ordinary and unparadoxical in that we can easily describe the circumstances under which one would say that the mind is dependent on the body. For example, "I'm very thirsty because I have not drunk enough water; as a result I can't concentrate on anything." Or, more reflectively, "I'm so thirsty that I cannot make myself concentrate no matter how I try; the spirit is willing but the flesh is weak," or again, perhaps in a class on neurophysiology, "I'm polydypsic because, as a consequence of my failure to replace body fluid loss, there has been a change in the osmotic gradient between the intracellular fluid in the osmoreceptors of my hypothalamus and the surrounding fluid, which change has caused these osmoreceptors to discharge and which discharge effects changes in the

---

[1] This conclusion is not dependent upon the truth of particular neurophysiological observations but upon the general relation between psychical and physiological events. What is significant is that psychical states appear to be correlated with physiological states.
[2] *The Blue and Brown Books* (New York, 1964), pp. 7-8.

cerebral cortex that are correlateable with my sensation of thirst; further, as a result of these areas being under such stimulation their activity cannot be synchronized in a manner requisite for highly discriminatory responses to complex stimuli."

But, second, there are trouble spots such as one's shock when one picks up a dead acquaintance's brain at autopsy and thinks, "Here is where he heard, remembered, etc." To a certain extent this may be the shock of thinking, "This is all the personality of my acquaintance was – this now congealed protoplasm and nothing more." The shock is the feeling of the floor giving out beneath the realm of human values, leaving one with *just* the relations of physical parts. But the floor has not fallen. One has rather suspended reference to man as a person. One's usual sphere of human values still remains intact; one is now involved in another sphere or dimension of thought.

These observations suggest a distinction between physical and mental strata of inquiry and appearance: [3] i) the domain of naturalistic experience – the domain of the natural sciences insofar as they exclude all reference to psyches, entelechies, etc. The body considered purely as a complex biochemical structure falls wholly within this domain and for our purposes is its paradigmatic example. ii) The domain in which one experiences the mental – the domain of the *Geisteswissenschaften*, including desires, wishes, motives, goals, and laws such as "humans have an inclination towards caring for their young." A paradigmatic example of this domain is the human mind and all it embraces as connatural to it: its plans, ends, loves, fears, works of art, tools, etc.

When one views the mind naturalistically, the result is a reduction of the mind to the body via an abstraction from all that is mental as such. Instead, for example, of thinking of the kindness and emotional sensitivity of one's acquaintance, one thinks, perhaps, of his limbic cortex, since one knows that in the context of neurophysiology, kindness and emotional sensitivity are *only* behavioral indices of limbic cortical activity, etc. Or if one wants to examine the domain of naturalistic inquiry to see what it reveals humanistically, one "internalizes" it in the sense of taking it as part of the world view of a certain epoch in the development of culture. Thus, the body as a particular neurophysiological structure (to which, in certain inquiries, one reduces the mind) is, as a theoretical scientific con-

---

[3] This analysis of experience into strata may be reminiscent of that given by Nicolai Hartmann. See, for example, *Neue Wege der Ontologie* (Stuttgart, 1949). Our analysis differs from Hartmann in providing i) a transcendental grounding of the categories and ii) a dialectical connection of the categories governing the physical and mental strata of reality.

struction, a historical production of mind, apart from any considerations of its objectivity. Complete or incomplete, accurate or inaccurate, it is an aspect of an epoch of the historical development of man's world view.[4]

But one can want more than a physical or historical reduction. One can lift up the brain at autopsy and wonder how it is possible to conceive of a unity of such diverse domains of significance. The question turns about a basic relation in meaning. Can the diversity of physical and mental reality be grasped in the unity of a concept? Moreover, can this conceptual unity itself be found in reality? The first question points to a project in thought – can one erect a conceptual scheme in terms of which mind and body can be understood to be related through a notional necessity? The second question points to a phenomenology – can an explication of the presented significance of mind and body reveal a notional unity? These questions are poles of an essential quandary – can nature be understood in terms of reason's requirement of conceptual unities and conceptual grounds? Otherwise, nature must be merely understood in terms of fortuitous relations. But the possibility of fortuitous relations points to the project of outlining a general notion of the relation of types of contingencies. One is returned to the problem of the notional unity of reality.

But neither the social nor natural sciences address themselves to delineating the notional necessity in the relation of mind and body. To ask for the intelligible basis of the relation of mind and body is not to ask merely for another psychophysical fact about their correlations and interdependencies. Rather, it is to ask what must be presupposed about the significance of mind and body and their relation in general in order for there to be the possibility of knowing psychophysical interrelations and interdependencies. This, of course, does not demean the enterprise of ascertaining naturalistically and culturally the relations of mind and body. But it does force us to engage in the special project of a categorial account.

Importantly, such an account has, in virtue of its phenomenological aspects, an implied thesis concerning appearance. Appearance refers to reality in all its richness, not a "basic" specially impoverished reality – a manifold of sense data. The world as we experience it is already a coherent whole furnished with objects, persons, tools – even social structures. The cup on the table does not appear as a mental construct, an

---

[4] See Husserl, for example, "Nature as a correlate, as that which is known at this stage, as the 'worldpicture' given by the science of such and such a time belongs within the cultural sciences, within history." *Husserliana*, Vol. 4, *Ideen II* (The Hague, 1952), pp. 392-393.

object fabricated out of sense impressions. It is already an object which has a certain purpose. True, one must learn its purpose. Indeed, the infant must learn to "recognize" it as an object. But these are empirical facts that are the subject matter of empirical psychology or at the most of very special areas in theoretical or rational psychology. In any case, for the adult the world appears as a rich connexus of well-formed objects and social relations in which one recognizes not only his brothers, sisters, friends, etc., but in which even certain idealized entities appear – for example, the family. That is, one recognizes social relations; they appear. If one does work up the manifold of appearance into a coherence, recognizable as a family, that would be an independent question. We start instead with the world in its richness, the world accessible to the normal adult. We accept it as a given fact and begin by explicating its features. This is to be countenanced since the claims made by the phenomenologist are open to universal scrutiny, as are all empirical claims. They do not begin with an intuition of the occult, but with an examination of the reality before us.

B. CATEGORIES AND CATEGORIAL ACCOUNTS

A categorial account of being seeks to reconstruct being according to indispensable notional unities. The "being" reconstructed is ambiguously both a pure concept and reality itself as given in experience. There is an ontological and a phenomenological pole to a categorial theory. The phenomenological pole is essential, it presents us with the problem at hand – being or a sphere of being. It gives us the reality that must be reconstructed, the starting point for the investigation. This starting point must be first recognized as a reality of a certain type before one can attempt to reconstruct in terms of a notional unity. This recognition of reality is phenomenological. Being as it appears to us will be inspected to discern the unities it presents. Of course, these unities will be themselves notional. A mind encounters only conceptual unities. Being as an object for a mind is always united according to patterns apprehended in concepts. It may be objected that one could not recognize conceptual patterns in nature unless they were already latent in the mind. This is surely in a sense correct. But one must not thereby confuse the order of discovery with the psychological or epistemological presupposition for discovery. We begin with the recognition of the presented notional structure of appearances, since a certain feature of appearance has excited our attention.

Moreover, in the absence of a phenomenological introduction, "category" designates only a general determination of the notion of being. It is an essential element in a general, fully determined notion of reality. As such, a category makes no claims concerning the actual existence of beings, it concerns only the general possibilities of being. As an element in the construction of the notion of reality, it is a necessary element in the structure of any actually existing, full reality of being. Any being not possessing some of these notional determinations would at the very least be unknowable. Still, "category" designates only the general possibility of structure, it does not assert the existence of an actual structure. How, then, is one to know if in fact the categories are instantiated if not through an examination of experience? Being may exist in a deficient form. Further, the problem of discovery arises – how is one to know what being is in order to give a categorial account or reconstruction of this notion? What is to suggest the general features of being if not experience of being?

A phenomenological discovery of the categories of being provides theory with a development out of that which was given prior to theory – experience – which is that which also presupposes a theory for its explanation. A phenomenological explication of experience secures the extrasystemic knowledge that the categories of being with which we are concerned are also the categories of being as it is for us. The instantiation of the categories is guaranteed, since we started from these instances.

That categorial accounts are also notional reconstructions of reality does not undercut their claim to be phenomenological. Notional reconstructions give conceptual models of reality; they articulate the concepts indispensable to a notion of "being." But the selection of concepts and their articulation is not arbitrary. The selection is dictated by experience and thereby involves a phenomenology. The articulation is phenomenological in an extended sense – it involves the discovery of the conceptual unities ingredient in the very presented significance of reality. It is peculiarly notional in that appearance is reflected upon in terms of reason's requirement of unity and explanation. One can say that phenomenology explicates or lays out the categorial structure of being, while a notional reconstruction explains or understands this categorial structure in terms of the unity of reason.

Until these issues are developed and others examined, any account of the idea of a "category" or a "categorial account" is provisional. But in this vein we propose to suggest the features of this crucial term "category."

1. *Categories*

A category is a note of significance integral to appearances in general or at least to a general domain of appearances.[5] Such a category is not merely general, it is necessary to the sense of an experience. It is not merely a notional structure, but a notional structure which occurs *in vivo* in appearance. Thus, a category as we shall use it refers not just to a notion, but to a notion found in concrete relation with the contingent possibilities of actual reality. If it were otherwise, it would be abstract and merely a category of thought or a category of being as possible, not a category of being as actual. A category of being, as we shall use it, is a general way in which appearance *is* for us.

Categories can have different ranges of universality; they are universal at certain levels or in certain domains: ideal, physical, mental, social, and religious. There are also categories given in any domain of experience. Thus, domains of experience are at least possible in which whole systems of categories could be absent (and for that reason the domains would be called "deficient modes of reality"). For example, a universe devoid of societies is logically possible in the trivial sense that it is not a formal logical contradiction. This, though, does not exclude "society" as a general category of being, for it is an essential characteristic of an entire domain of appearance. Such a category we will term constitutive.

A constitutive category is to be distinguished from categories that constitute the particularity of objects and from those that constitute their mode of existence. The first will be termed categories of particularity and the second categories of presence.[6] In a sense, all categories are constitutive, for they are the notional structure of being. But the categories of particularity and presence (the categories of individuation) are distinguished from constitutive categories proper in not constituting the significance of a particular domain of being. We will explore these distinctions now as an introduction to our later investigations.

---

[5] We employ the term appearance as a constant reminder that we have started from experience. As will become quite clear, the use of "appearance" and "being for us" does not imply that there is a noumenon or being-in-itself irreconcilably other than and alien to phenomenal reality. It serves only to emphasize that we have begun with the presented significance of reality, not with assumed definitions of reality or self-evident truths. Our point of departure is thus phenomenological.

[6] This distinction was suggested by Stephan Körner in a colloquy paper, "The Impossibility of Transcendental Deductions," given at the University of Texas, Austin, 6 January 1967, and by C. S. Peirce's distinction of firstness, secondness, and thirdness, though the distinctions of both these authors differ from mine.

a. Constitutive categories

A constitutive category is a most general albeit indeterminate significance of an object. It is constitutive of a domain of appearance by delimiting and characterizing it. For example, the category of physical objects establishes a very general domain of objects. The significance of being a physical object is a general category of unity; it is the note of permanence in terms of which the appearances of merely constrained, non-intentional events occur. (This notion will be developed in Chapter II A1 and 2). In short, a constitutive category provides the prime note of meaning within a general domain of appearance.

One must distinguish categories, which constitute a domain of reality, from eidetic types and notes of reality found within a domain. The universality of the first constitutes the significance of a domain of reality. For example, space and time constitute the domain of natural objects; the category of brute givenness constitutes the presence of any object within that domain. Further, the categories of mind and body delimit more determinate domains: the biological world and the cultural world. Within these worlds there appear eidetic types which are merely essences. An essence is the eidetic sense of a variety of appearance but does not constitute a general domain of appearances. Instead, it characterizes and delimits what appears as a mere variant of a general mode of being. The word "type" has an important ambiguity. On the one hand, it can indicate an embracing, constitutive feature of reality: the state is a type of reality as a category. On the other hand, it can merely indicate an essence: a government by concurrent majority is one type of state, while the "state" is a category constitutive of the domain of political reality.

In short, essences occur within a domain of being which categories determine. They explain the various presentations of a particular type of appearance; they are the intelligible structure of that type and are in this restricted sense principles of explanation. Categories, in contrast, are true ontological principles. They are not only determinations of thought found in being but determinations of being grasped in thought. Such principles are not merely attributed to appearance via a reflection on appearance. They serve as primary determinators of the notion of being. In this respect the category "organism" is distinguished from the essence "dog." The first is a general determination of thought; it is a concept in terms of which the notion "physical body" receives a further and richer determination. The essence "dog" serves to make a certain collection of appearances intelligible but is hardly a principle of explanation invoked to give a general account of being.

b. *Categories of particularization*

Physical bodies are not encountered as concepts. They are given to us as *particular* bodies; indeed, minds are given to us as *particular* minds. Thus, we can distinguish categories constitutive of objects within a specific domain of appearance from those that particularize the objects so constituted. The former are the general significance of any object within a region of appearance prescinded from the object's note of particularity. It is in terms of the latter categories that particular objects are constituted.

Of course, the categories of particularity are also principles in terms of which a domain of being is constituted. But they constitute this domain only insofar as they constitute the mode of particularization of its objects. For example, spatiotemporal sensible appearances are always in this respect particular.[7] This refers us to the presented structures of a certain dimension of appearance: the particularity of its contents in space, time and history. When the categories of particularization are considered pictorially, they are continua in which objects are particularizable (e.g., Kant's forms of intuition). Considered notionally, they are categories of space, time, and history.

c. *Categories of presence*

Further, individuality is more than particularity. It has, besides the note of being a one over against a many, the sense of presence. The category of presence is the category of the mode of existence of an object — mental, physical, mathematical, fictive. It is the category of givenness concerning which one can distinguish the brute givenness of physical objects, the necessitated givenness of mathematical objects, the capricious givenness of fictive objects, the appresentative givenness of other minds, and the self-conscious givenness of one's own mind. The categories of givenness are so intimately related to the categories of particularization that it is difficult to consider them in isolation. But if one recognizes the sense of an object as one over against many possible similar objects of the same type, then one can contrast it with the mere sense of an object as present, apart from its being one among many. The latter is the note of presence.

The category of presence, as indicating a mode of givenness, has a patently phenomenological character. However, it is also notional in that it indicates the mode of existence. The brute givenness of a physical

---

[7] Of course, space and time are also constitutive categories – the most indeterminate constitutive categories of being. See G. W. F. Hegel, *Sämtliche Werke*, Vol. 9, *Die Naturphilosophie* (Stuttgart, 1965), #254-259.

object is its contingent existence in contrast to the givenness of mathematical objects which expresses necessary existence. Thus the categories of presence bear a similarity to Kant's categories of modality in that they concern the relation of the subject to the object.⁸ But unlike Kant's categories they do further determine the objects concerned. The brute givenness of physical objects is integral to their significance, though this note of significance contains a special reference to subjects. It can in this case be compared to the notion of *Anders-sein* at the beginning of Hegel's *Naturphilosophie* – the pure note of physical reality. But in addition it includes the note of "hereness," of "haecceitas." ⁹

One can speak then of the mere presence of a particular object, which particularity is realized in the object's being a particular object of a certain type. For example, the mere brute givenness of an object is realized as the hereness of a particular spatiotemporal object, which particularity is realized in the object's constitution as a physical object. Thus, one has three strata of categories: categories of presence, categories of particularity, and constitutive categories.

d. Contingent qualities

Finally, contingent qualities, qualities that do not constitute an object's sense of being an instance of a general type of appearance, further constitute the particularity of the object. Though any one of them is contingent to an object, it is necessary that some one of them be there. Though a particular shape is contingent for a physical object, some shape is necessary; in order to be a particular object of any general type, an object must have particular, contingent qualities. Thus, the contingent qualities are delineated a priori, though only in general, via the constitutive attributes of the object (as *some* shape is necessarily required for being an extended object). These contingent qualities are how the general type of an object realizes presence and particularity. The constitutive categories are presented in a particular quality of a general type. Contingent qualities, then, do not constitute a separate stratum of categories. They are the constitutive categories of an object, not considered abstractly, but presented concretely in experience; they are the constitutive cate-

---

⁸ See Immanuel Kant, *Critique of Pure Reason*, A 218=B 255, A 219=B 266.
⁹ See, for example, Husserl's discussion of *haecceitas* in *Husserliana*, Vol. 4, *Ideen II*, #64. This is also similar to the presentness of C. S. Peirce in the sense of "whatever is such as it is positively and regardless of ought else." (*Collected Papers of Charles S. Peirce* [Cambridge, Mass., 1965], 5.44.) But ours is not a general category of quality as Peirce's category is. Indeed, it may bear closer resemblance to Peirce's category of struggle or secondness.

gories as possessing the necessary moments of presence and particularity. In phenomenological description of the appearance of mind and body we must then always attend to the relation of the individuating as well as the constitutive categories in order to avoid an abstract description of reality. To do otherwise is abstractly to isolate dimensions of the eidetic features of appearances which are notionally interrelated in experience. In the experience of the physical world generals are not given as isolated but are given in brute particulars.

In fine, a constitutive category is a general and indispensable determination of the notion of being. Its reality component is usually expressed in its relation to categories of particularization and presence. A category is the conceptual unity of a plurality of beings (an exception to this: God). The truth of a category, though, is found in the notional unity provided by its meaning. The notion of a "state," for example, is the conceptual unity of what would otherwise be merely an aggregate of interests, social associations, etc. It is the intelligibility of a plurality of appearances, and in virtue of this intelligibility the plurality is one. A category is a unique and encompassing unity of the significance of being.

2. Mind and body – a categorial relation

There is a domain of appearance which is characterized as that of physical objects. The least determinate, yet most essential, qualification of an object in this domain is that it is a body. This is the case even if one would wish to be more specific in the use of the term "body" and indicate only organically organized bodies. In doing so, one would set the limits of an extensive domain of appearance constituted by the sense "organic body." *Pari passu* mind constitutes a domain of appearance. We will term both mind and body categories. In doing so we refer to them as general notes of significance that characterize the most general kind of appearance within a domain of experience. This brings us, as we have noted, to looking for a unique form of relation. We are interested in the basic connection between these constitutive categories. More precisely, we are looking for an affinity of meanings, since we are concerned about the possibility of understanding the relation between such diverse domains of appearance. We are interested in comprehending a categorial relation.

A categorial relation is not one relation among others, but is the ground of the possibility of a domain of relations. In the case of mind and body, the search for their categorial relation is the search for the least determinate but most essential characteristic of their relations. It is least determinate in that it is not any particular, empirically presented relation; it is essential in that it is eidetic in any particular empirical relation of mind and body. It is the note of significance that cannot be absent without the absence of that domain of relations. That is, we want to understand the relation of mind and body in general.

We can anticipate the results of our investigation and indicate a model for such relations – what we will term a founding-founded structure. The term "founding-founded" indicates that one category (the founded category) presupposes another (the founding category) for its (the founding category's) full significance. The structure is identified by the fact that the founded category contains the founding category as an essential moment of its significance and not the reverse. The term "moment" thus indicates an essential element of a category. The moment itself is categorial but its significance is no longer understood independently, rather within the structure of the category of which it is a moment. Importantly, this notion of a founding-founded structure of significance embraces both 1) the notion of the stepwise constitution of the sense of an appearance in successively richer strata of significance and 2) the articulation of categories in a system that progresses from the lesser to the more determinate. Again the phenomenological and notional poles of a category are evident. It is because of this dimorphism that our analysis will be diphasic. Here, though, it is enough if we realize the possibility of a certain structure of appearance and be acquainted with the terminology that could be used for its explication. The establishment of its existence must wait until later (i.e., Chapter II and IV).

C. PROGRAMS OF INVESTIGATION

The analysis of the categorial relation of mind and body presupposes a mode of inquiry that is neither natural nor social scientific. It will not focus on particular facts or general regularities but on the categorial structure of being. The project falls into two levels, one phenomenological, one transcendental. The first is merely explicative. It lays before us the categorial structure of appearance. As such it serves to introduce us to the philosophical problem at hand. The second is explanatory. It focuses on the notional or dialectical necessity that underlies the presented categorial structure of appearance.

1. Phenomenology

> ... Be it understood, then, that what we have to do, as students of phenomenology, is simply to open our mental eyes and look well at the phenomenon and say what are the characteristics that are never wanting in it, whether that phenomenon be something that outward experience forces upon our attention, or whether it be the wildest of dreams, or whether it be the most abstract and general of the conclusions of science.
> Peirce, *Collected Papers of C. S. Peirce,* 5.41

We must first be clear about the categorial structure of mind and body before we can look for the dialectical necessity that may be its basis. Thus, we begin phenomenologically. Phenomenology is the project of describing our experience of the world. It is not a part of phenomenological reflection to determine, deductively or otherwise, the real significance of the material before us. Appearance is observed as it is given immediately and what is given mediately in the immediate appearance is so noted. For example, a physical object has the sense of having a back side, even though that side is not immediately given. Or to take a more extreme case, the position of a subatomic particle has the sense of being determined by a past event; what is unclear is whether the determination is strict or statistical. The openness to either possibility is part of the sense of subatomic particles. To decide one way or the other is not a matter for phenomenological reflection, but a theoretical judgment of physics. Phenomenology notes the sense of appearance: what it presents i) immediately and ii) mediately as further possibilities for experience. One brackets the theoretical interpretations we give to reality in observing it, so as clearly to *note* both appearance and our experiencing, including our interests and theorizing. Obviously, we cannot free ourselves from the stream of experience, step outside of it, and reflect. But we can deliberately attempt to note appearance without introducing ourselves *qua* observer. The result is a presentation of the fabric of reality which can be termed being-for-us.[10]

---

[10] One might ask what sort of reduction does this entail and how it compares with Husserl's phenomenological reduction. To avoid extraneous historical questions, this question will be left open. But if one must, one could say that our project entails a reduction to its presented basis of any reality claim, not immediately made by appearance. That is, insofar as appearance presents itself as a coherent connexus, with the sense of being independent of any particular observer, its reality is for us and is not bracketed. What is bracketed are claims that transcend the presented fabric of appearance. Thus the possibility of an ontology is not first excluded and then reintroduced. The doctrine of appearance, the structure of being for us, awaits an ontological account.

The *us* for whom being is, the *we* who reflect, is any empirical subject who assumes the stance of a reflective observer in the fashion indicated (above) as requisite for phenomenology. As *any* observer, he can as well also be *none* in particular, but rather a viewpoint that we try to reach, free of idiosyncratic prejudices. As such a stance, "we" is the eidos ego, the ego considered in terms of what is eidetic to its significance as, in this case, a reflector upon appearance.

This stance is similar to Kant's notion of experience. Experience, for Kant, is anonymous; the transcendental unity of apperception is not anyone's in particular, but is eidetic to everyone's experience. It is the unity constitutive of experience; the name of the experiencer is superfluous. Similarly, the observer of the eidetic structures of appearance is the eidos ego, which is to say the particular phenomenologist is no longer considered as particular. An eidetic insight does not belong to anyone in particular. It is reflection upon the world from a "theoretical" stance in the sense that one does not necessarily attribute the reflection to a Unique Eidos Ego. Rather, it is a stance we all assume qua phenomenologists of the eidetic in experience.

This "we," this stance, interestingly is similar to Hegel's "we" of the *Phenomenology*. Both we and Hegel are attempting to bracket our involvement in experience so as to reflect upon appearance without the assumptions integral to the natural attitude we possess in the world of everyday life. "There thus enters into the movement of consciousness a moment of being in itself and for us, which is not presented to the consciousness within the grip of experience." [11] The we is the phenomenological observer who has bracketed his involvement in experience in order to grasp experience and his involvement with it.[12]

a. *Eidetic phenomenology*

As Peirce has pointed out, "The business of phenomenology is to draw up a catalogue of categories and prove its sufficiency and freedom from

---

[11] *Werke,* Vol. 2, *Phänomenologie des Geistes,* p. 80.

[12] The similarity of our phenomenology to that of Edmund Husserl's depends upon the extent to which his critique of the naive apodicticity of phenomenology (see *Husserliana,* Vol. 1, *Cartesianische Meditationen,* #63) would lead to a category of the eidos ego as absolute mind. That is, it is a question of moving from an aggregate of egos, *de facto* sharing in an intersubjectivity, to a category of complete experience and full grounding of the categories of experience. The difference between the eidetic stance of any of an aggregate of egos and the categorial apprehension of the notion of a fully objective stance is the difference between the phenomenology of Husserl and that of Hegel and ours. At least, this is the case unless Husserl moves to a category of objectivity which transcends the level of particular subjects as such. Husserl may be hinting at such a move in *Ideen II,* #64, when he speaks of the "Absolutheit des Geistes."

redundancies, to make out the characteristics of each category, and to show the relations of each to the others." [13] This is its business, for if one is to explain appearance one must deal with its general structures. It does not seem too farfetched that if one is to do this, then one must attempt to isolate these structures through examining appearance. Further, it does not seem unwise to do this via an inspection. After all, appearance is open to any observer; it is objective. This will be our point of departure. In beginning our investigation of mind-body, we will reflect upon its phenomenal structure and hope to find those aspects necessarily inherent in any presentation of mind and body. These aspects must be seen to be those features of the mind-body relation which constitute its sense as such. They will have to show their necessity in the fact that their absence entails the absence of an appearance with the sense: mind-body relation. We will term these features eidetic: the necessary notes of a type or category of appearance.

Eidetic phenomenological description cannot then strictly speaking be abstraction, for one is not interested in isolating structures from appearance. In order to explain and justify the articulation of categories of mind and body we must start with being as it is presented to us. This is where our questioning arises. The first step is thus an explicative attending to an appearance; one is not separating out an essence but pointing to what in the appearance is essential or eidetic. One can test such assertions by asking, "would the appearance remain the same type of appearance if feature A, etc., were altered?" To find a feature whose absence changes the type is to find an eidetic feature of that type of appearance. One must recognize a change of type as one does a change of colors. One can see that the introduction of hue does not cause a change identifiable within the achromatic color series, but instead produces a chromatic color. Similarly, for example, without rationality man is not man: a variation of this feature of the type of appearance changes its sense. Eidetic description functions as an ostensive definition of categorial features of appearance. It points them out *in concreto*. It says, see the type; see that changing element A changes its entire sense. The ostentation is ideational. This does not viciously beg the question concerning the eidetic. Instead, it returns us to what is in question. One shows that an onion is osmatic by offering it again and again, "Here, smell it." The only difference is that when one grasps an eidetic feature, there is no need to repeat the inspection. What has been pointed out is a feature of reality integral to the possibility of a domain or type of experience. Consequently, the

[13] *Collected Papers of Charles S. Peirce*, 5.43.

feature is discloseable in all future repetitions of this experience, since it is a condition of this experience. Further repetition is thus theoretically redundant. One may find this insufficient. But one must remember that all knowledge of the world presumes that we can recognize its structures. Nothing more is asked here than that one recognize the generalities in appearance.[14]

b. *A phenomenological requirement: an adequate account*

Eidetic phenomenology presumes that one has adequate access to appearance. This is a problem which is prior, in the order of knowledge, to the problems of eidetic phenomenology. First, there is the presupposition that the natural sciences have unearthed whatever appearances are needed for an eidetic phenomenology and which are not readily available through everyday experience. This is crucial only within special restricted and particular domains of appearance; the general notes of appearance are present in all appearance and do not wait upon natural science for their revelation. Second, there is the fundamental requirement that we have adequate everyday experience. This again only holds of those features that are not immediately required in the appearance of any object whatsoever. Though the first requirement is absorbed by the second, we will treat each separately in order specifically to consider the problems of adequate scientific investigation and its relation to a phenomenology of mind-body.

i. *Adequate scientific knowledge*

In the project of a phenomenology of mind-body, one is not concerned with a) a factually exhaustive investigation, or b) even an investigation that will bring forward all empirical laws. One is rather concerned with an investigation that will bring forward general types of appearance. There is, for example, the general fact that the mind is dependent upon a particular organ whose properties are correlated with psychical events. The radical nature of the dependency of mind and body, implicit in the very sense of a spatiotemporally intuiting finite mind, is nevertheless hidden. Science's role in such a case is a contingent and merely heuristic one. It provides the study of the mind-body problem with: a) an explicit presentation of what was only implicit before – the radical dependency of the mind on the physical states of a quantity of organized matter, b)

---

[14] Naturally one is not expected to accept this without further examination. This choice of method will, at least in part, have to be vindicated by the course of the investigation.

a further clarification of what was previously presented explicitly yet not in full clarity – psychical effects of drugs, etc., which, though previously known, are presented in greater completeness and lucidity. Science furnishes us with experiences that further elucidate what was intrinsic to, though covert in, the appearances with which we were already acquainted.

Thus, we must often wait for science to draw our attention to features of appearance that later are obvious. For example, if Aristotle had had more extensive neurophysiological information, he would not have asserted that the mind cannot "even have an organ like the sensitive faculty." [15] He would have been acquainted with the specific dependencies of higher mental activities upon portions of the cerebral cortex. Further examination would have explicitly revealed the dependency of the mind upon the body, bluntly showing the world to have a character other than the one that had been hastily assumed. The character of the mind-body relation exhibited by science is, in this case, one that was there all along, only more easily overlooked in the course of everyday experience. And it is not unexpected that this dependency should be often overlooked in one's (onesided) marvelling over the universality of mind and its transcendence of nature.

Scientific knowledge would seem to have problems uniquely its own, since a domain of which it treats can never be directly presented. Yet, though inferred entities and regularities offer certain epistemological difficulties (i.e., are they merely hypothetical constructs?, etc.), they can be treated phenomenologically in a manner analogous to ordinary experience. The problem becomes one of describing what sort of world it must be in order to imply a certain type of inferred entities and regularities. In any case, though one can conceive of cases, perhaps in the physics of subatomic particles, where science could introduce a level of reality with novel types of structures, physiology does not offer us such a special and novel type of appearance. Physiology simply makes more explicit what was already directly presented.

For example, the eye has (among other things) the sense or significance of being the locus of one's receiving visual sensations. If the eye is closed, one stops seeing the outside world. Covering up one's eyes and not one's ears impedes vision; one puts corrective lenses on or in front of one's eyes, not one's tongue. This is already sufficient to establish the eyes' significance as being i) one's mode of access to an entire sphere of

---

[15] *De Anima*, in *The Basic Works of Aristotle*, tr. Richard McKeon (New York, 1941), 429a26-27.

appearance. ii) It also presents the dependency of one's perception upon the physical status of the perceiving organ. The physical status of the perceiving organ is only emphasized in scientific experience. Further examination presents the eye as transducer of photic energy into electrochemical energy, transliterating the pattern of photic energy into a patterned transmission of electrochemical energy. A simple example of this is given in the fact that rhythmically blinking one's eyes in bright light causes a corresponding rhythmic alternation of alpha and asynchronous higher frequency lower voltage electroencephalographic waves. The eye's significance as a receiver of visual sensations is observed to have the sense of being radically dependent on its transducing light into neural impulses (i.e., if this transduction is interrupted, sight fails). These experiences emphasize the eye's sense of being both "my visual organ" and a "physical object with certain remarkable properties." But these experiences add nothing that is essentially new.

Moreover, these descriptions, though they concern theory infected experiences (photic energy, electric energy, neural tissue), point primarily and directly to experience and only secondarily to theoretical constructions of it. After all, one measures photic energy differently from electrical energy, neural tissue looks and stains and otherwise reacts differently from other tissues, etc. All these are senses directly presented in appearance, and the general conclusions do not involve inferred entities or hypothetical constructs. Indeed, they are an immediate extension of everyday experience. We can therefore turn to the problem of adequate experience in general, with the conclusion that experiences of mind-body in neurophysiology do not provide a special difficulty for phenomenology.

ii. Adequate experience

The problem of seeing what is already, though only implicitly, given, is to be solved by prudence, diligence, labor, and intelligence. These remedies are practical, and not theoretical. Practically, there is often little difficulty in a general survey of a restricted domain of appearance. The capital theoretical question is altogether different: by what criteria does one judge that experience has presented to him all the needed types of appearance? The question is serious, since an inadequate inspection of appearance can lead either i) to a distorted, confused account whose incomplete set of types refuses to cohere in a complete unity, or ii) to an account whose elements are forced into a unity that does not correspond to the notional structure of the world. Appearance, though, does not immediately present a criterion of its completeness. Because this is lacking, one can have phenomenologically only the ideal of adequate inspection, never its certified fulfillment.

Confronted with this difficulty we will attempt to certify adequacy in terms of an account's encompassment of being. Encompassment is crucial, for an account can be onesided; it can be adequate to one dimension of reality without taking account of the rest. An account of the types of naturalistic appearance could be adequate as such, but insufficient in not encompassing all relevant strata of appearance. A solution to this problem of balancing our treatment of the dimension of the mind-body relation may in turn solve the question of adequate experience. It may allow us to anticipate a priori what the general structure of an adequate account would be. As a result, though we lack a phenomenological criterion for adequacy, we will attempt to find an ontological criterion which is, in this case, the same as a criterion for the proper articulation of the findings of eidetic phenomenology.

c. *Eidetic phenomenology as ontology*

Eidetic phenomenology is an exhibition of the structure of being; it is descriptive ontology.[16] Thus we recognize only Kant's strictly negative use of the noumenon.[17] It is affirmed only as a restriction of knowledge to the realm of possible experience so that the categories of experience are the only categories there can be. These are the categories of being insofar as we can refer to it.[18] Reality, contingency, particularity, *haecceitas*, etc., are given in experience and there is no need to invoke a positive noumenon to explain them. They explain themselves in forming a unity of experience. The objectivity of the world is understood as the possibility of objective experience by a subject. Experiences by particular subjects are partial realizations of this objectivity of appearance. There is no obstinate other, no matter, no receptacle or primal amorphous *res extensa* that must be a non-categorial principle of explication, much less a co-justifying ground alongside reason.

Ontology is the exhibition of the categories of being-for-us, including

---

[16] The identification of eidetic or transcendental phenomenology with ontology is not a novelty but was indeed suggested by Husserl. See *Husserliana*, Vol. 1, *Cartesianische Meditationen*, #59 and 64; Vol. 8, *Erste Philosophie* Zweiter Teil, pp. 212-228. Moreover, it could be argued that Hegel considered his *Phänomenologie des Geistes* to be at least a provisional ontology.

[17] See *Critique of Pure Reason*, A 254-255=B 310.

[18] As will become evident, since the notion of a noumenon is abandoned, the notion of experience is enriched to include experience of persons. We will introduce a category of subject, not as a category of transcendent being, but as a category of appearance.

INTRODUCTION 21

both those that concern the constitution of diverse types of objects and those that constitute their interrelation.

Ontology describes the fabric of significance ingredient in appearance. After all, the necessary condition for the possibility of the appearance of anything is that it appear in some way and not in others and that it be recognizable as such. Things do not nor can they just appear devoid of any significance, for they would have no relation to anything, not even to any experience; that is, they could not even appear. Appearance or experience cannot exist in the absence of distinguishable categories or types. If it so existed, it would be the maximum of monotony – a pure being of being indistinguishable from non-being.[19] A world in any significant sense must be constituted of being of diverse types or categories. Appearance must have general unities. Without such unities, a world could not even appear as a set of unrelated points. There would be nothing in terms of which the points could appear to be related, for relations (including unrelatedness) are instances of categorial unity. Diverse elements presuppose a unity in relation to which their diversity can appear, in terms of which the diversity is part of a domain of being. In the absence of a general or categorial unity there would be no domains of appearance, for nothing would provide a note of coherence. A thing that could not have any general relation could not even appear; the absolutely isolated individual is as such impossible.

Thus insofar as things are a part of a single cosmos that appears to us, they must have interrelations. Furthermore, the relations cannot be arbitrary, but must follow from the sense of the relata, or there would be no possibility of a coherent world appearing. This holds as well for the relation of categories. Without a minimum of notional affinity and thus unity, there would be no community among strata of meaning, among spheres of actions, etc. They would not form *one* universe. To be a universe each element must have the sense of being relatable in certain ways to the other elements.[20] A cosmos that presented no conceptual unity could never be for us anything but a name for nothing. For an appearance to be an appearance, for a cosmos to be *a* cosmos, it must have a unity of elements apprehendable in thought. Diverse categories can cohere in one world only if they possess some affinity. This can be expressed as a general principle for ontology: being must always be under-

[19] Such would be the being with which Hegel's *Logik* begins – the most abstract and indeterminate notion of being.
[20] That the general structure of appearance must be the affinity of diverse elements in a unity was recognized by Kant in his maxims of manifoldness, affinity, and unity. See *Critique of Pure Reason*, A 662=B 690, as well as A 651=B 679, etc.

stood in terms of both its unity and its diversity. Further, this structure of unity in difference must be *recognized* if we are to construct an ontology that is not onesided. This principle requires that we not onesidedly fix on one of its dimensions of an ontological relation, but encompass in our account both its elements of unity and diversity, e.g., its physical and psychical reality. This requirement is termed ontological *encompassment*, or the principle of unity or identity in difference.[21]

This principle is a regulative ideal of balance or encompassment which can act as a criterion of adequate phenomenological investigation. The principle requires that one not exclude one dimension of experience under the pretext of complete explanation. Thus, an account of mind and body which excluded their unity and asserted only their dichotomy would a priori be inadequate, indeed onesided. *Pari passu*, a monism would be inadequate since it ignored the distinction of mind and body. The principle of identity-in-difference exhibits a priori the necessity of such a diversity. As a general structure of appearance it serves as a guide for adequacy in the explication of particular spheres of appearance.

2. Transcendental ontology

Eidetic phenomenology reveals for mind and body both their unity and their diversity as distinct categories of appearance. But as purely descriptive explication, it cannot explain or justify these "facts." It answers the categorial *quid facti*, but not the *quid juris*. The answer to the *quid juris*, the justification of the unity of these two categories in terms of their own sense or significance, entails a turn to transcendental reflection in which one explains the affinity between categories implicit in them as a *conditio sine qua non* of their being contained in a notional unity. It is a meta-reflection which introduces no new content, but shows the ground of the necessity of structures of appearance already known via eidetic description. Thus it begins with the notional unities and affinities discovered by phenomenology and explains them. In doing so transcendental philosophy presents the structure of thought about the structures of appearance. The categorial structure is explained by exhibiting the dialectic implicit in this structure: the reasons why one category is incomplete and requires another for completion or requires another as its moment. The reflection is in terms of the goal of explanation; one justifies the relation of the categories as a progression towards more complete

---

[21] This principle is a primary maxim in Hegel's philosophy. See, for example, *Werke*, Vol. 8, *System der Philosophie 1: die Logik*, #88 (4).

explanation. The goal of full explanation operates as a principle of justification through which the categories and categorial relations revealed by phenomenology become "principled" terms of a rational explanation. Consequently, we contrast explication, which is descriptive or phenomenological, with explanation, which is truly notional or transcendental.

Our use of "transcendental" is not unlike Kant's, for like him we are not directly concerned with particular objects, but with the mode of necessary knowledge about objects.[22] But unlike Kant and indeed unlike Husserl, we are not just concerned with the necessary conditions or eidetic types of appearance, but with their notional unity. One could say that we are interested in specifying the unity of the categories in the transcendental unity of apperception. Or, *pace* Husserl, we are interested in developing a critique of the "naivete of apodicticity." [23] We wish to grasp the eidetic or transcendental unity of the categories, of the eidetic types basic to appearance.

But, when Kant identified the articulation of the unity of knowledge (thought about thought, a priori knowledge about a priori knowledge) with pure reason, he attributed *merely* subjective validity to the conclusions of reason. Since the conclusions of transcendental philosophy do not apply directly to objects, but rather to the categories of appearance, it would be a category mistake to think of transcendental or dialectical affinities as immediately presented affinities between objects. They are rather the ways in which we understand the notional affinity between the categories of appearance. But these affinities need not for that reason be any the less objective or intersubjective than the categories exhibited by eidetic phenomenology.

Our use of the term "transcendental" thus manifests important aspects of the Copernican revolution begun by Kant and completed by Hegel. First, the grounding of categories is no longer sought in the objects they refer to. Categories are determinations of appearance but they will now be understood as stages of explanation ordered in terms of an ultimate theoretical viewpoint. The term "transcendental" thus indicates that reality is not ultimately to be understood in terms of transcendent objects, but rather in terms of principles of explanations grounded in the unity of thought. This use of "transcendental" is quite compatible with Hegel's

---

[22] I. Kant, *Critique of Pure Reason*, A 11-12=B 25. Also, "A transcendental principle is that through which the universal determination, under which alone things in general can become objects for our knowledge, is presented a priori." *Kants Werke*, Vol. 5, *Kritik der Urtheilskraft* (Berlin, 1913), p. 181.
[23] Husserl suggests a critique of the naivete of apodicticity; see *Cartesianische Meditationen*, # 63.

enterprise of giving a systematic unity to the categories through their grounding in Absolute Mind. Indeed, Hegel can be considered the first to have completed a systematic turn from objects and the transcendent, to mind as the higher truth of objects; he is the first to have completed the turn to the transcendental.[24]

The transcendental turn is not esoteric. It is simply the project of giving a full notional unity to the categories of appearance. The unity is assured through the notional character of the categories. In that the categories are the general determinations of being recognized in thought, the categories are themselves determinations of thought. As determinations of thought it is not overly optimistic to presume that we can analyse their notional affinities and thus their unity in thought. The subject as the ground of objective experience insures a notional unity for the categories in and through which the world is presented to us. This assertion will be examined in Chapter IV. Here in anticipation we can suggest two synergistic ways of understanding the notional unity of the categories: i) as the unity of reality found to be the unity of thought, ii) as the focus of objective experience.

i) Either the unity of reality is to be found in being apart from thought or it is to be found in being for thought. The first case is that of a noumenal grounding of the unity of reality, the second is the grounding of appearance in the unity of thought.[25] But unity is not to be found beyond the phenomenal world in a positive noumenon. First, there is no way of reasoning from phenomena to noumena, since by their very meaning they have no "community" and therefore their interrelation is surd. Second, if one could make reference to a noumenon, the noumenon would remain unknowable and mysterious and useless in a reasoned account of categories. The positive noumenon is always a surd irrelevant to the categories of appearance. Consequently, either the categories of reality and their unity are other than thought and thus being is for us superbly irrelevant, or the categories of thought are the categories of being. Either being is consonant with thought or it becomes a noumenon towards which we mutely and blindly gesture. There is no third alternative. To assert that the unity of appearance is in part alien to thought is to assert that a certain coherence of being is in fact incoherent, that that

---

[24] See Klaus Hartmann, "On Taking the Transcendental Turn," in *The Review of Metaphysics*, XX (December, 1966), pp. 223-249.

[25] This so far has strong Kantian tones; the unity of the categories can be understood as the transcendental unity of apperception. All categories of appearance have a unity since they are uniteable in an "I think." See Kant, *Critique of Pure Reason*, Transcendental Deduction, B edition.

which is grasped in thought is not in fact understandable through its principles. But insofar as coherence is recognized it is *eo ipso* recognized in thought. Indeed, incoherence insofar as it is recognized is a type of coherence. In short, the unity of being is found to be understood in terms of its unity for thought. The next step is to attempt to explain this unity in terms of the notional affinities that the categories possess in virtue of their recognition by thought. In particular, we can attempt to explain this unity in terms of a central demand or goal of reason.

ii) To understand the subjectivity or idiosyncrasy of any experience one must make reference to a nonidiosyncratic viewpoint. This can be phrased phenomenologically. Objective reality is given to us in experience. The world of appearance is not *merely* our construction, it is not merely subjective. It presents itself as objective reality; it is not understood and justified from the idiosyncratic viewpoint of a particular subject. Objective reality has the sense for us of being that which would be appropriated by a subject freed of its idiosyncratic limitations. But most importantly, the categories of being do not have the sense of being merely my thoughts, but a thought structure apprehendable by any mind. Subjectivity only makes sense insofar as one can appeal to a standard of objectivity. Otherwise, each subjective experience constitutes a world unto itself and is an objective apprehension of that world. To recognize an experience as subjective implies that a world of experience has a possible unity independent of any particular subject's realization of that unity. The nomological regularity of the world and the notional unity of its categories are posited as an ideal independent of any particular subject.

Further, in positing this objective viewpoint one posits a coincidence of being and thought. The theoretical point of the actualization of full objective truth is the theoretical point of the full manifestation of being. Being in itself is what would be known at this point of full objective truth, it is what would be understood from the theoretical stance of all objective knowledge, including the categorization of that which is not itself conceptual (i.e., contingency, feeling, etc.).[26] It is through positing the possibility of such a full realization of being in thought that we understand our present viewpoints to be idiosyncratic, onesided, and incomplete, and yet viewpoints of the same universe. As philosophers, we explicitly refer to the point of coincidence between thought and being in

---

[26] Or to quote C. S. Peirce, "Finally, as what anything really is, is what it may finally come to be known to be in the ideal state of complete information, so that reality depends on the ultimate decision of the community..." *Collected Papers*, 5.316.

order to ground our categories which otherwise would neither refer to reality nor be adequate to the sense of objectivity.[27] This point of coincidence expresses the singularity of objectivity and truth. Further, it is then possible, in terms of this coincidence, to order categories and justify them as different stages of the objectivity of reality. Each expresses an aspect of the coincidence in that each is integral to the significance of a domain of being. Each could then be articulated in respect of its compass of being, the richness of the domain of appearance which it constitutes. Each category would then represent a further and more determinate notion of being ranging from the least determinate to the most determinate notion, thought fully embracing being. The final category would be the notional articulation of all the previous categories, the full notional unity of all the categorial determinations of being. It would be the full categorial comprehension or explanation of being by thought (i.e., the point of coincidence of being and thought). The goal of full categorial explanation or comprehension of being thus enjoys a unique position as a principle of reason. It expresses a signal relation between reason and the categories of appearance: being's grounding in thought.

We will employ this principle to justify the articulation of the categories, to elaborate a dialectic of meanings. The dialectic will express the movement of thought from a less encompassing domain of concepts and appearance to a more encompassing one. What was recognized phenomenologically as a *de facto* stratification of domains of ontological richness will then be understood as a rational nisus towards a theoretical viewpoint that would be fully inclusive and fully differentiated. Particular dialectical movements consist in the apprehension of categorially new standpoints from which the previous domain of appearance is understood as a moment of the newer, richer domain.[28] The new domain is the emergence of a kind of being which presents itself as the notionally

---

[27] C. S. Peirce makes a similar point. "And what do we mean by the real? It is a conception which we must first have had when we discovered that there was an unreal, an illusion; that is, when we first corrected ourselves. Now the distinction for which alone this fact logically called, was between an *ens* relative to private inward determinations, to the negations belonging to idiosyncrasy, and an *ens* such as would stand in the long run. The real, then, is that which, sooner or later, information and reasoning would finally result in, and which is therefore independent of the vagaries of me and you." *Collected Papers of C. S. Peirce*, 5.311.

[28] Thus, for example, when Hegel asserts that "spirit has for us as its *presupposition* Nature of which it is the truth and thus of which it is the Absolute First," he is asserting a dialectical relationship between ontological categories: the notion of a world is integral to the content of the notion of spirit, while the further significance of the world is to be found in the category of spirit. *Werke*, Vol. 10, *Philosophie des Geistes*, # 381.

natural revision and completion of the lower domain.[29] Transcendental ontology attempts thus to relate the categories in terms of a dialectic of explanation or comprehension of reality. Our present investigation is at most only a segment of such a complete ontology.

D. LEGITIMACY OF THIS INVESTIGATION

Insofar as one can grasp the possibility of explicating categories and relations of categories, one can at least provisionally entertain the possibility of this investigation. To assert that one cannot delineate the rules for the relations of domains of meaning or sense presupposes that one has grounds for demonstrating the impossibility of such an attempt. It is not sufficient merely to assert that such an attempt can lead to confusion. This is, for example, Wittgenstein's position when he asserts that all that can be sensibly established concerning the relation of mind and body is a correlation of "a train of images, organic sensations, or on the other hand of a train of the various visual, tactual and muscular experiences ... in writing or speaking a sentence," and the experience of "seeing his brain work." [30] The attempt to understand the nature of the relation beyond the merely empirical correlation, he dismisses as senseless: "Both these phenomena could correctly be called 'expressions of thought'; and the question 'where is the thought itself?' had better, in order to prevent confusion, be rejected as nonsensical." [31]

If this is meant to suggest that questions concerning the categorial basis of the relation of mind *and* body are necessarily misguided, then Wittgenstein is ignoring a question presented in experience. As pointed out above (See A 1 and 2), there are experiences of this relation which call for a grammar relating the domains. Indeed, this investigation is necessary if we are to know the canons of thought concerning the relation of mind and body. As our knowledge of the physical dependence of our mental life upon our nervous system increases, we become more acutely aware of the difficulty of speaking correctly of their relation. Nor are we apt to accept an expedient silence in order to avoid the occasion of a possible confusion. Rather, we are pressed to answer; reason seeks an explanation. We must find our language and thus our thought wanting until we are able to account for these categories in conceptual terms.

---

[29] The revisionary nature of the dialectic is described in J. N. Findlay, *The Discipline of the Cave* (New York, 1966), especially Chapter III.
[30] *The Blue and Brown Books*, p. 8.
[31] *Ibid.*, p. 8.

CHAPTER II

A PHENOMENOLOGY OF MIND AND BODY

Phenomenology is a program of descriptive ontology. It avoids all theorizing which transcends explication of appearances. To limn this character of our program we will refer to phenomenologically explicated features of being as "senses" or "significances." The senses or significances of appearances are the presented meaning of being-for-us. The general constitutive significances of appearances are the categories of appearance. Insofar as they are intellectually apprehended as notes of meaning, they are notional. They are presented as apprehendable by reason.[1]

The primary phenomenologically disclosed categorial relation is the founding-founded structure of categories. It must be assiduously distinguished from the dialectical relation of categories. The first is a nonexplanatory, descriptive explication of categorial relations: one category (the lower or founding category) is discovered to be necessary to the significance of another category (the higher or founded category). The necessity of the lower category is not understood notionally in terms of a goal of full explanation. The relation cannot afford more than a hint by the founding category of the founded category. It cannot actually indicate the higher stratum without transcending its own significance. Such a transcendence of a lower category by a higher is a dialectical progression in meaning necessitated by the posited stance of full explanation (see Chapter I C2). In contrast, the founding-founded relation is a *de facto* characteristic of categories. Phenomenology does not explain this progression of meaning, but notes the resultant founding-founded structure.

With this in mind we turn our attention to mind and body in order

---

[1] The features of appearance are apprehended as rational because of the grounding of appearance in reason. See the discussion of objective reality in Chapter I C2. This will be further developed in Chapter IV A2.

to reflect upon their appearance and explicate each category and their relation.

A. EXPERIENCE OF MIND-BODY

1. *A phenomenological description: preliminary*

What is it to walk across the room? One can see the opposite wall and *think* "I would like to be *there*." One can then *desire* and *will* to be *there*. But that is not enough. Before one can have that location, one must begin to move. This involves a mechanical process of changing locations and is describable in a naturalistic manner (i.e., devoid of any references to persons or minds). The change of location can be viewed merely as a naturalistic phenomenon, a mass moving. In ordinary life this is presented in moments of exhaustion when one must, so to speak, drag oneself across the room. The body's physical properties loom as hindrances and must be coped with. In such a state one realizes in tedium the mechanical processes that must be effected in order to change location. One places one foot after the other doing what must be done in order to change location. One must attend perhaps to the positioning of one's legs in order to avoid toys the children have scattered on the floor. One's body becomes another object to be mastered; but yet not quite, since short of death one's title to his body is never completely lost.

An opposition, a diversity has begun to emerge counterposing mind and body. The thinking, desiring, and willing that we have noted are each "of"; they are directed towards a sense or significance. They are a "thinking of (being at the other side of the room)," a "desiring of (being at the other side of the room)," a "willing of (being at the other side of the room)." The "moving," though, simply as such, has only the sense of being a changing of place, a process of ceasing to be here and becoming there. The moving itself is not an attention toward any thing, but a process of passing through locations. Of course the movement can become an object of intention: "I want (to be moving)." One attends to the significance of moving and desires it in some manner. But the moving itself is not a way (except metaphorically) of attending to or intending anything; it is a way one realizes a project in the world. Nor do "movings" attend to or intend anything. They have the presented sense of a mechanical spatiotemporal event.[2]

[2] The term "mechanical" is to signify physical as opposed to mental processes and to stress their non-psychical character. The mechanical is that which is "caused by, resulting from, or relating to, a process that involves purely a physical change." *Webster's Seventh New Collegiate Dictionary*, 1966, p. 525.

To walk across the room thus has more than mechanical significance: the shift in the location of a mass. I, a person, walk across the room. There is a place here for senses or significances besides the purely mechanical or physical. My desires, intentions, thoughts, volitions, perceptions, sensations, etc., are all personal senses or significances. They are all oriented around a personal identity pole. They are my, or your, or his desires, intentions, thoughts, etc. They are not oriented around physical objects as such, around poles of merely mechanical physiochemical activity. The sense of desirer, intender, thinker, willer, etc., is not applicable to, is not possessable by what appears as merely physical, as a pole of merely mechanical activity. A chair, a turbine, a car, a centrifuge, a river, can only metaphorically possess such a sense. But the human body has the remarkable ability of being a physical object capable of more than merely physical significance.

The being of humans is thus complex. It appears in two domains of significance, each with its own types of coherence. On the one hand, when my friend talks to me and tells me of his thoughts, desires, volitions, etc. (or when I turn to my own), they have a purely personal sense in that they are considerable apart from any physical embodiment. On the other hand, my friend's body (or my own) can be considered apart from his thoughts when I consider it merely as an object for anatomy or physiology. My friend as mine (granted an abstraction) and my friend as a physical body (likewise an abstraction) contrast. As a mind he is a unity of attendings to or intendings; he appears as my friend, the same person through time. The features of his personality are interlaced by habitual ways of feeling, thinking, acting, etc.; he appears as a personality, intendings and intendings of intendings woven together through time in a mosaic whose design is determined by the qualities or kinds of intendings. That is, the features of his personality are interlaced in ways describable in mental terms, apart from whatever physical causality might be involved. Perceptions and sensations are not just felt, wanted, or not wanted; they are so in particular ways that constitute the features of *a* personality.

A mind presents itself in wantings, desires, thoughts, etc., that assume some habitual pattern. But importantly, these thoughts, wantings, and desires are not unified mechanically in a personality as muscles are unified on a skeletal system. Their relations are via the intending of their sense. For example, I regret my desire to have X and resolve in the future to desire Y even though distracted by the desirable sensations of X. Or, overwhelmed by a compulsion for desirable sensation X, I neglect Y, etc.

These elements (desires, resolves, sensations, etc.) are related by one's intending their senses, their significances, not mechanically through forces.

On the other hand, in sharp contrast, if I consider my friend's body merely as a physical object, I am confronted with a system of elements of various masses, united by sundry forces: adhesive, cohesive, electrical, van der Waals, etc. The elements are related to each other by position, mass, quantity of force, and quantity of reaction to attractive or repulsive forces. Even if one would contend that attractive and repulsive forces among and between electrons and protons, etc., were limiting cases of intentionality, this would be beside the point. The physiological level of examination is concerned only with the mechanical and chemical properties of the body, not with a fabric of intendings. Similarly, in physics one is concerned with quantitative forces in space, not with qualitative gradations of intentions. When one views the living human body naturalistically, one views it only as a very complex assemblage of space occupying elements, of particular masses related to each other not in intentions but through laws extraneous to the sense or significance that they present as desiderata, *intenta*, etc.

But within this contrast a unity emerges. It is in and through my body that I experience and know the world. Perceptions require physical contact with the world – most obviously so in the case of touch. Physical relations provide the basis for sensations and desires. When one burns his hand he initiates a physical sequence of events; but this physical process "results" in a strong desire to withdraw one's hand from the heat. Further, the physical unity of the body is a basis for the unity of the mind. It is the mind's unified access to the world. On the other hand, the mind provides a unity for the body – it is in consciousness that the body is experienced as one, as I in the world. We will first analyse the sense of mental and physical unity integral to each domain of experience. Then we will turn to the relation between these unities.

2. *Identity poles of appearance*

"Identity pole" is used to refer to the presented unity of an appearance through time and through change. An appearance that maintains its sense as one appearance through time and change has a unity to its diverse changes that can be termed its "identity pole." This unity is not asserted to be a substance, a *res substratum*, but a presented sense of permanence through time in relation to which changing properties acquire the sense of changing properties, not just incomprehensible sequences of appearances. The identity pole is the necessary condition of the objective experience of both objects and subjects as such. "Permanence," as Kant explained, "is thus a necessary condition under which alone appearances are determinable as things or objects in a possible experience." [3] Taken as an eidetic phenomenological description, it characterizes the objectivity of both things and persons. For an appearance to have the character of being an appearance of a person or an object, it must have the sense of being a focus of senses or significances, or being a point of permanence in terms of which various characteristics have the sense of being the characteristics *of* an object or person. In particular, it should be noted that this permanence (as Kant failed to point out) is a presented requisite for the experience of a subject; it is the sense in terms of which the appearance coalesces as the appearance of a subject. Persons are given to us as the *same person* day after day; I am given to myself as the same person as I was yesterday, only with certain changes. The identity pole is the presented sense required for this coherent experience of objects and subjects, in short, for a world of things and persons. It is the point of objective reference within the appearance of objects and subjects. It is a category of appearance.

But in this coherence of experience the objective pole of identity and the subjective pole contrast. On one hand the mind is presented as an identity pole of thinkings, willings, valuings, perceivings, feelings, and actions (i.e., in the sense of personal actions, *Handlungen*). It has the sense of being the identity pole of all that is peculiarly personal, of all that belongs to persons not as physical objects, but as subjects. It is the identity pole of processes that are directed to their respective objects, that is, which are intentional in character (e.g., thinkings of X, willings of X, etc.). These processes are peculiarly related to a meaning (i.e., to a significance or sense); they are thinkings of a cognizable-sense X, or willings of a wantable-sense X, or perceivings of a perceivable-sense X, or feelings of a sensible-sense X, or actions that are processes having the purpose of accomplishing X.[4]

The term "identity pole" refers to the mind having the sense of being a focus of unity for these processes, even if it may only be an axis for a

---

[3] *Critique of Pure Reason,* tr. Norman Kemp Smith (London, 1964), A 189=B 232.
[4] One can, following Edmund Husserl, distinguish the act of intending from the sense intended as the noesis versus the noema. This distinction of "thinking" versus the "thought" is not a separation of discrete elements but of two poles of consciousness understood in its broadest sense as the field or domain of significances. See *Husserliana,* Vol. 3, *Ideen I,* #87-96.

bundle of perceptions.⁵ Free floating thinkings, willings, actions, etc., are not presented to us, but rather A's thinkings, A's willings, etc. The grammar of words for mental processes is such that they require a substantive to give them their station or place (e.g., one does not have perceivings that are no one's perceivings). Further, the choice of the term "identity pole" is intended to avoid the implication that there is a mental substance, a *res cogitans* underlying mental processes, in which they inhere.⁶ Such a substance would go beyond what we wish to examine here – namely, the appearance, mind.⁷ The term "identity pole" indicates that the mind appears as the point of unity, the pole of origin and possession of the processes mentioned above. It is an intelligible note ingredient in experience, not a thing, much less a thing in itself.

On the other hand, the body, as a material object, has the sense of being an identity pole of physiological, biochemical, electrical and other varieties of physical occurrences and processes. It is in short the identity pole of occurrences and processes that appear or are viewed as impersonal. The processes of which it is an identity pole are not directed to their object, are not intentional. These processes and occurrences do not have the character of a relatedness to a meaning (i.e., a significance or sense). All these physical occurrences, both mechanical and chemical, take place externally to the meanings (senses or significances) of the appearances they concern. Physiological processes, unlike perceptions or purposeful actions, are not directed towards a meaning or sense perceived, desired, etc., at least when viewed naturalistically (i.e., when viewed prescinded from their relation to persons). They transpire externally to or prescinded from the meaning or sense of the objects concerned.

Subjective and objective identity poles of experience are our first eidetic isolates.⁸ First, as we have noted, the absence of an identity pole

---

⁵ Hume's failure to account for the sense of personal identity may be traced to his faulty description of the structure of reality. Appearance is not merely composed of isolated units of significance (impressions and ideas) and their conjunction through association. Rather, structures of unity, themselves presented in experience, are integral to reality. Compare, for example, Hume's attempt to account for the personal identity in the *Treatise of Human Nature,* Book I, part IV, section vi, and his admission of failure in the appendix.

⁶ Hegel, it should be noted, drew attention to the fallacy of reifying the subject in his criticism of Kant's treatment of the paralogisms. See *Sämtliche Werke,* Vol. 8, *System der Philosophie 1, die Logik,* #47, pp. 137-139; Vol. 19, *Geschichte der Philosophie 3,* pp. 577-579.

⁷ Terming the mind an appearance is a recognition that minds are presented to us in experience, a fact that Kant tried vainly to deny.

⁸ The notion of subject and object poles of identity are discussed at length by Husserl. See *Husserliana,* Vol. 1, *Cartesianische Meditationen,* especially #27, 30, 31, 32, and Vol. 4, *Ideen II,* #22, 23, 25.

is the absence of the sense which constitutes an object for consciousness. Willings of X, seeings of X, etc., which are no one's seeing, etc., are not possible components of a domain of experience. The same for motions, attractions, etc., that are not the motions and attractions belonging to anything. Second, for a motion of a body to be the intention *of* a new location *by* a body is to bring the body beyond the limits of its being able to appear as a mere physical body. It must also be a mind. Or for a mind not to be qualified by intendings of noemata, but rather to be merely the subject of processes (e.g., attractions, repulsions, motions) is for it likewise to be beyond the bounds of possible appearance as a mind. It is, rather, a body. Thus, the unity of an appearance is integral to its significance: a subject pole is not an object pole. The unity is not separate from the diversity which it unifies, nor shall we prescind from the diversity. We will be attending to the concrete unity mind and the concrete unity body, making special cognizance of the unity ingredient in their appearance.[9]

Consequently, we have isolated two distinct and eidetic structures of appearance: a) minds, the subject poles of appearance with their intentions; b) bodies, the object poles of appearance with their processes. Each is a distinct and general structure of significance whose absence would entail the absence of a stratum or domain of appearance. In the absence of the significance "subject pole" there could not be subjects nor therefore a domain of mental reality. *Pari passu* in the absence of the significance "object pole" there would not be a domain of physical reality. Thus, not only are these categories, they are constitutive categories.

3. *Mind and the Ego*

Mind and body are distinguished in the contrast of the domain of physical processes and the domain of intentions. The sense of permanence in

---

[9] One can attempt the nominalist ploy of denying any actual unity to the successive moments of an object or subject, and claim instead that the unity is merely virtual. In such a case one may claim that the unity is only the result of custom, as did Hume, or due to the fabric of inner time consciousness, habituations, and dispositions, as does Aron Gurwitsch. But insofar as the unity of mind is presented in appearance, whatever the psychological cause, it has the logical or phenomenological claim of being the unity of the appearance. Indeed, if feelings have the sense of being someone's feelings, then one has encountered the notion of the ego as the unity ingredient in minds. It is our allegation that such unity is encountered. Nor is this merely an assertion, since as we have noted in Chapter IC diversity is inexplicable without the notion of unity. Appearance is an identity in difference in which constitutive categories are poles of identity. Thus, *pace* Gurwitsch, the ego is not "nothing other than the concrete totality of the disposition and action it supports"; it is the unity of this totality. See A. Gurwitsch, *Studies in Phenomenology and Psychology* (Evanston, Ill., 1966), p. 298. An ego is not a thing, but neither is a note of unity.

each domain has been termed an identity pole; it is the focus of unity for the diversity in each case. It is the general note of unity eidetic to a stratum of appearance. As such, object pole and subject pole, or mind and body, are broad and relatively indeterminate categories. But now we must distinguish these from more determinate categories with which these might be confused. This will require referring to some notions which could only properly be introduced later: organic body, I and person. This subsection is intended to forestall questions concerning who (the mind, the self, the ego, etc.) the possessor of "my" body is.

"Organic body" is the least problematic, for it can be treated as the category of body characterized by a specific type of unity: it is a universalizing mechanism. The self-contained unity integral to the notion of a complete, automatic, and self-controlling mechanism is absent in and different from the level and sort of unity possessed by stones, rivers, etc. Moreover, it can contain the complexity requisite to respond in a varied yet highly structured way to the vicissitudes of the environment; its range of operation is of a different level than atoms and molecules.

The notion of I or self is more involved. It is a further specification and determination of mind inessential to the basic relation of the categories of mind and body. Mind is a unified domain of consciousness. Consciousness is used most broadly to indicate any mode of intending a significant note. It can be awareness devoid of any self-awareness; a feeling of a sensum but not a knowledge of that feeling. In that case, the unity of that feeling within a particular mind is recognizable only by another mind or the same mind at a different time. That is, there may be good reasons for speaking about unconscious feelings (e.g., either the pressure of my shirt against my back before I self-consciously attended to it or the feelings of an opossum). If one can characterize such cases, then one must distinguish different types of "egos" or unities of consciousness. Most generally, the ego would not necessarily be an "I," or a self-consciousness. Rather, it is first the general note of unity of a number of intentions; the note of unity presupposed in the recognition of a mind. In terms of this ego the activity of a mind has the sense of being the series of actions of *one* mind. Thus the ego is the subject pole of identity, the permanence of a mind which we identified above.

The "of" expresses the most impoverished sense of mineness. One can speak of one's body and its sense of being "mine" and refer only to its being categorially presupposed by a mind. Higher, richer, and more interesting senses of mineness involve, at least in part, self-consciousness of the relation of mind and body. For the richer senses, there must be recognition of one's own unity as one's unity over against the unity of objects (and subjects). The ego as I is thus the self as self-conscious. Though the determination is more complex, this affords a basic characterization of the self as a further determination of the mind. By recognizing the self as the determinate realization of the mind by the mind, we preserve mind as the fundamental category of consciousness. As long as we proceed no further in our description of the self, we are free to move from describing the category of mind to describing its relation to *my* body. That is, one can seek to understand the mineness of the body in terms of the general relation of mind and body recognizing that a richer epiphany of mineness is to be found in the relation of a more determinate category of mind (e.g., self or I) and body.

We will thus distinguish the ego, the unity of mind, from the I or Self. In our usage every mind will have an ego, but not every mind an "I." Insofar as we will speak of the I as "I experience my body as mine," no more will be meant than: it is noted that the body enters into the field of consciousness united around a particular ego and that I know this since I the observer am this ego. That is, the body is lived in by mind. We are not introducing the specific strata of significance integral to the "I" as such. Reference to the "I" is not avoided. It is less problematic to note the restricted use we make of this notion or significance than systematically to isolate a uniformly indeterminate stratum of experience from the rich experience of reality possessed by man. We are, after all, human phenomenologists. The "I," the referent of "my," is the observer of the unity of mind and body. He knows that he is in some way identical with his mind and his body. For our purposes it is enough to note that he is so identical since the I is a further determination of the significance "ego." This is but to recognize the range of the notion of "subjective identity pole." The full determinate significance of "I" is a matter for detailed explication, but need not be borne by this investigation. In that we are interested primarily in the dialectical relation of mind and body, and only incidentally in the problem of the significance "my body as mine," we need not analyse the various layers of significance ingredient in "mine" and "I."

It is enough to note that we are dealing with an identity pole of intentions that can be aware of itself as such and thus reflectively apprehend its own significance. When we point out the significance of the mineness of the body, we are thus indicating its most indeterminate sense. The body appears as the presence of mind in the world. But since this mind also has the sense of being me, we shall simply note: the body appears as I in the world. The use of "I" stresses that this significance is given firsthand in experience, and does not depend on content unique to the "I."

The same is the case in our use of "man" or "person." In describing experience we speak of "man's experience," "my experience," and refer to the experience of a "person." With proper reservations this should also be unproblematic. "Person" can designate both the unified appearance of mind and body, and the notion of an ethical agent. We will for the most part be referring to the presented unity of mind and body in which what belongs to the body also belongs to the mind (e.g., "I have a spasm in the muscle of my legs") without intending to specify this unity beyond the categorial relation of mind and body. In attending only to this we avoid the question of a mundane person's significance as an ethical agent. Such higher strata of significance are simply ignored. We do not, though, exclude them. Indeed, they may of necessity (dialectical necessity) grow out of an undetermined category of mind. In the last chapter we summarily examined this possibility. Below we intend to give a complete description of mind-body only in the sense of outlining its most general features. Therefore, when we speak of determinate structures one is warned to consider them only in their most general significance.

4. Naturalistic experience of bodies

Natural science has brought into relief senses of appearances that were until recently obscure, or to some extent merely implicit. Indeed, the natural scientist and the physician often create a new life-world for themselves by prescinding from the dimension of persons in order to focus with undistracted attention upon the purely physically objective. The immediate result is not a theoretical construction, but the isolation of a segment or a dimension of the world within the natural attitude.[10] In it, deprived of their relation to excluded dimensions of sense, certain structures of significance are given with unaccustomed force, and are therefore recognized for the first time. For example, the human body, considered not as a living body but as a merely physical body, may be recognized as being an extremely versatile, complex, adaptive and discriminative machine. I may notice how automatically it responds even

---

[10] The "natural attitude" designates the philosophically unreflective attitude of everyday life in which objects, etc., are uncritically accepted as real, independent entities. In short, it is the world unreconstructed by a metaphysical theory. See *Husserliana*, Vol. 3, *Ideen I*, #27-32.

when both the stimulus and response are complex. Indeed, the medical student learns to observe his own life processes in order to notice the dimension of the purely physical and mechanical; the purely mechanical dimension presents itself as having a full reality and legitimate claim to being a concrete stratum of appearance.[11] For example, the placing of an object in the lower portion of the pharynx precipitates a complex movement of muscles which we term "swallowing." Knocked off balance the human body responds by initiating complex muscle movements that re-position limbs.[12] These reflexes are presented to us as occurring without any conscious involvement, and their mechanistic sense is exhausted when we display all the parts interrelated in the execution of these occurrences. The body is found to have the sense of being a machine.

One realizes that the experience of one's body as a physical object is not a limiting case confronted in "unusual" circumstances such as iatric or traumatic anesthesia. It is a limiting case indeed, but one confronted with frequency. Not only is anesthesia a more frequent experience given the increased incidence of its use by dentists, but one has everyday confrontations with the automaticity of his reflexes. One bangs his patella and the knee jerks. One touches poison ivy and there is a noticeable biochemical reaction, one drinks a sufficient quantity of beer and there is a rise in urine output, one drinks too much coffee and the heartbeat becomes irregular, one runs and the rate of respiration increases uncontrollably. Indeed, one discovers a dimension of the body in which the sense of personal action and involvement is inappropriate (e.g., one does not have the sense of being personally involved in the bleaching of his hair by the sun). The body is presented in this naturalistic attitude as a physical object, albeit an extraordinarily complex one.

In this the body presents itself to me as an other, as an object among others to be felt, seen, and acted on. Indeed, its otherness can appear in the very acts of feeling, seeing, acting, as, for example, when one notices that his leg has gone to sleep, when one's vision blurs, when one with physical difficulty effects an action in the world. The body is presented as *something* to be *lived in*, to be enlivened. One massages the limb with poor circulation, desiring to feel in it and act with it. One's lived body is

[11] Our use of "naturalistic" experience describes this non-cultural stratum of significance which Husserl identified as the correlate of the naturalistic attitude. See *Husserliana*, Vol. 4, *Ideen II*, p. 179.

[12] There are, for example, primitive pathways for the response involving only the semicircular canals, the vestibular nuclei, the reticular formation, and the nerves of the reticulospinal and lateral vestibulospinal tracts – all of which are components of the brain concerned with automatic, "non-conscious" processes. This remains true even if such reflexes always involve mediation through the palio- and archi-cerebellum.

always potentially opposable to one's mind as merely a body to be enlivened. It is *that* which one enlivens. In this experience, a "distance" between the mind and the lived body presents itself: the body is objectifiable in a way in which a mind, a person, is not. A mind is never merely an object, but is also a subject. The "distance" between the mind and the lived body arises out of the distance between the sense of an identity pole of intendings and the sense of an identity pole of physical processes. They are two contrasting categories that constitute two distinct domains of appearance.

The inappropriateness of personal significance in appearances of the merely mechanical is an epiphany of this distance. But there is also a special relation. Personal significances or senses can be founded on certain physical appearances and not on others. The physical appearance of a friend is not of merely physical significance, but at once also possesses personal meanings: my friend the happy, obese philosopher, etc. Such significances or senses are not possessable by other sorts of physical objects: chairs, centrifuges, etc. These are observations about how appearances appear and what sense they have of not being able to appear otherwise and still be the same type of appearances. They suggest a founding-founded relation between the categories of body and mind, as well as the identification of a category of body distinct from physical objects in general.

5. *My body as my location*

Human bodies do not possess merely physical significance, but appear as lived bodies.[13] Such a body does not possess personal significance in the way that a mind does, yet neither does it possess such significance in the metaphorical sense that a mere body would. It is, as it were, the point of meeting between the mind and body, the point where the significance of both join. It is the bearer of personal significances. It is where I see, think, feel, act. The body is the limit of nearness.[14] It is only in relation to my body that something is near, but far enough away so as to have a spatial distance from me. The body is the standard of nearness and therefore is

---

[13] The term "lived body" is introduced to designate the body as the body of a living human being. It is to be contrasted with both inanimate physical bodies, as well as with the human corpse or the human body prescinded from any significance not generally presented by physical objects. This distinction is to some degree realized in the contrast between the German words *Leib* and *Körper*.
[14] See Husserl's treatment of the lived body as a center of orientation, *Husserliana*, Vol. 4, *Ideen II*, #41.

itself not near, but rather I. This sense of hereness is presented via the fields of kinesthetic and tactual sensation and is approximately coterminous with them. The immediately felt sense of bodily location, kinesthesia, forms the schema of sensation in terms of which there are localized sensations.[15] As a field of localized sensations the body acquires the further sense of being my center of orientation in the world. As this center, it is my unified physical and sensitive presence in and through environmental changes, etc. It is I in the world. Or otherwise put, the body must be the localization of sensation if it is to have the significance of embodying mind. A mind that was embodied but only intellectually aware of this embodiment (i.e., a mind which knew of its dependence upon the central nervous system, but occupied itself with deriving mathematical theorems, not with sensing the world) would have to recognize itself as a deficient mode and limiting case of embodiment. It would realize only the least degree of a rich spectrum of significance; its experience of the world would be restricted to the experience of the dependence of its cognition upon the state of its central nervous system. But even as deficient it is still the experience of embodiment. Through this restricted possibility of experience, the body has the significance of being one's presence in the world; it fulfills the notion of sentient existence in the world.

This presence, though, is not uniform. Instead, certain organs and organ systems stand out as special epiphanies of the functions of the mind within the world.[16] That is, one's "hereness" is given via his body, but especially via the parts in which one is preeminently conscious. For example, because the eyes are in the head and serve as prime coordinators of kinesthesias and tactual sensations, etc., the sense of measuring "here" from the head is fairly explicit. Further, the sense of the relative importance of organs for conscious life (which has its roots in experience afford-

---

[15] Husserl, *Husserliana*, Vol. 4, *Ideen II*, #35-42.
[16] One may be hesitant to pursue this phenomenologically, since past failures (e.g., Plato's suggestion that lower consciousness has a special seat in the liver, *Timaeus*, 71b) suggest that this project involves a theoretical reconstruction of appearance, and thus is heavily infected with the science of the day. Therefore, it is useful to distinguish a proper domain for phenomenological enquiry. a) In respect of the amplification of our scientific understanding of phenomena and their relation via proposed laws, forces, etc., phenomenology can only describe what sort of phenomena can lead to what sort of possible theories in general. In that case, phenomenology is occupied not with the description of the appearances as such. Instead, it produces a phenomenology of science. b) But on the other hand, science can amplify experience by exploring new ways of acquaintance with the world. Such amplification of experience can then allow further and novel phenomenological description. It is our intention to proceed in this latter fashion by describing the experiences that further acquaintance with our bodies has provided. The relation between science and phenomenology has already been touched upon in Chapter I C1bi.

ed principally by natural science) is the most irrevocable foundation of measuring nearness from the head. This is recognized through acquaintance with one's own organs – a process that does not require a theoretical reconstruction of appearance and which is indeed an extension of ordinary experience.

As one becomes acquainted with what it means to be a finite, mortal, and composite being, the organs of the body are differentiated in respect of the sense of being me. They present a gradation of mineness, constituted by grades of what is essential to my continual conscious mundane existence. Even is ordinary life we have the notion of our central nervous system as the locus of our thinking and viewing the world. As we become more acquainted with persons having organs replaced, we will have this sense presented more clearly. The result will be to recognize a distance between myself and all replaceable organs; a unique sense of the nervous system will be revealed. The nervous system cannot be replaced and leave me intact. Rather, one experiences radical dependence upon the nervous system for sensation, perception, and thought. This is a natural extension of such experiences as a) I live only by continued physical processes of certain sorts: eating, breathing, etc., b) physical damage will or will not destroy a person as a conscious being, depending on which organs are injured and how severely. These experiences are an expression of the fact that I can change my location vis-a-vis things outside of me, but I never change my relation to myself (a useful tautology: I am where I am). When extended to varying one's relation to organ systems, one notices that one cannot vary one's relation to the organ system that is the embodiment (or the physical analogue) of mental life: the central nervous system. This is an eidetic fact since it delimits the possibility of being in the world. To donate the central nervous system *in toto* for transplantation would be tantamount to replacing all other organs; to donate certain parts would be to destroy one's personality. That is, I have the sense of being a composite existent that can be decomposed to a certain extent before ceasing to live; the sense of life in this world is not the sense of being a simple indivisible being. And further, a particular part of the body emerges, through experience, with the sense of being pre-eminently the location of consciousness.[17]

The significance of the nervous system as my signal locus is ingredient in the more encompassing significance of the whole body as my existence in the world. And this latter sense of the body is not founded just upon

---

[17] This theme will be explored further in Section A8 below.

the body's sense of being my locus of self-consciousness and sensation. Rather it requires in addition the sense of the body as active in the world. Even a purely passive mind must at least have the sense of being a spectator and thus engaged in the activity of being conscious in a world. Yet to develop this sense is to go beyond this limiting case of activity and enter into the notion of being truly active. That is, the mind's tendencies towards involvement in the world are realized when one passes from the mind as mere observer to the mind as observing agent. But the mind engaged in praxis involves a further enrichment of embodiment. The mind is no longer related to its objects as to independent entities to be sensed and experienced. In praxis objects are appropriated and made connatural to mind. They are not just known but rendered into means for goals, made into products of design. In doing so the mind itself is further integrated into the world and the human body appears more forcefully as the bridge to full participation in the events of the world. My body appears as that from whence and through which I operate upon the world. It is the unique vehicle of my agency, for my thoughts do not act nakedly upon the world. Rather in and through the body one finds oneself in the midst of and a partaker in the processes of nature. The body is the locus from which the mind transforms and imprints the fact of reality with structures conducive to its purposes.

In terms of the body and its location some actions are possible and others excluded. The location of the body determines at least in part what goals are relevant, what plans can be effected; projects are envisaged in terms of one's body, one's most primordial means for contact with the world.[18] Consequently, one can refer to the body as the mind's fundamental possession or tool. This mode of possession is unique.[19] A mind can be deprived of other tools and possessions, and yet in this impoverishment, even as a slave, it remains in the world. On the other hand, the body is a necessary condition for having tools and possessions. Tools exist only insofar as a mind physically enters into a domain of praxis: they are an extension of the physical reality of a mind. Still, the having of a body is analogous to the possession of physical objects. One can, for example, employ one's body as a tool – one can hammer with one's hand, inflate balloons with one's lungs, one can even sell parts of his body – his hair for wigs, a kidney for a transplantation. Yet a unique sense re-

---

[18] See Alfred Schutz's analysis of "Relevance," *Collected Papers I, II, III* (The Hague, 1967, 1964, 1966).

[19] This point is nicely explained by R. M. Zaner in *The Problem of Embodiment* (The Hague, 1964).

mains – the body is one's first possession. This possession cannot be diminished without essentially compromising one's presence in the world. Just as one cannot give away his nervous system without abandoning life in this world, one cannot have his musculature paralysed without abandoning the possibility of mundane action.[20] That is, the sense of the mind as agent in the world is preeminently associated with the musculoskeletal system. It is the fundamental location of any initiation of agency in the world. Conjoined with the nervous system, it forms a unity of experience and response within a spatiotemporal world. This unity is an outline of the significance of a special category of physical body: the organized body that can be the embodiment and thus the location of mind.[21]

Further, minds are also differentiated spatially one over against the other in virtue of being located by a body. The body appears as the mind's vantage point for perception and action. I perceive and act upon the world in bits and pieces. The world is not given to me in one unified experience, nor acted upon within a pure volition. Its constituents are given in a continuum of adumbrations and affected through a succession of actions. That is, the world has the sense of having cognitively and volitionally opaque elements that are reciprocally interrelated at any one moment. Any perception is (prescinding from temporal succession) only one adumbration of an object which could have been replaced by a different adumbration, had my orientation to that object been otherwise, just as every action would have its effects altered by a change in location and vantage point. This opacity and its indefinite openness to further partial perspectives and actions is integral to the sense of the complexity of physical objects. My perception of them and action upon them is a perception of and action upon parts, one outside of the other – each part contemporaneously and differentially excluding my perception of and effect upon other parts. Thus, I have a particular perspective in my perception of the world. But as such an infinity of other perspectives is possible. The world is a connexus of possible perspectives. One is localized and thus particularized spatially in virtue of his vantage point for perception and action. Other vantage points appear as possible locations for

---

[20] With the musculoskeletal system we include the motor cortex and the efferent nervous system. Progress has been made in developing mechanical prostheses that can be directed by action potentials from muscles still intact. One can recognize the possibility of developing prostheses directly attachable to efferent nerves.

[21] In the future the term "body" will, unless otherwise indicated, refer to the special class of physical objects that carry on the physical analogues of mental activity, e.g., the intricate response of the nervous system to specific stimuli which is the physical analogue of "recognition."

the embodiment of other minds. One is one particular mind vis-a-vis other minds with other perspectival coordinates.

In addition, I never perceive nor act upon the world instantaneously, but rather in temporal adumbrations, in a temporal series where every moment is after a previous moment and before a succeeding moment in which the processes of the present will have their consequences. Further, every moment has the sense of a temporal extension; time is not presented as composed of indivisible simple moments. Even the apprehension of non-spatial entities (e.g., geometrical theorems) is never all at once, for every thought can be analysed in a series of moments. All my consciousness is thus localized in time. It exists in the world in one age rather than another. The mind is consciousness in a particular time; it is particularized vis-a-vis its predecessors and successors. Only in relation to contemporaries is the particularization primarily spatial.[22] Embodiment is, thus, not just spatial but temporal. My body is when, not only where, I am. The body is when I see, feel, think, and act. It locates me in time and orients time to me. Events are in the distant or recent past, in the present or in the future through my body's relation to the process of the world's development.[23] The time of my body is the limit of nowness; through my body I am in immediate contact with the now of the world. Or otherwise put, the now of my body is the primordial sense of the present time of the world.

Again, the central nervous system has the most univocal sense of being my location. In being where I am, its temporal phases are the temporal phases of my embodiment. They are when I experience and act upon the world. For example, my being-now-in-the-world ceases with the serious dysfunction of my central nervous system. The subjective experience of time is dependent upon my embodiment. Further, it is only in terms of the central nervous system that consciousness can be strictly correlated with the world. This correlation of the now of consciousness with the time of the body is an identification in that all that appears for the objective assessment of the time of mental events are neural processes (or what is subsequent upon them, verbal communications). It is in terms of these neural processes that consciousness has a strict locus in objective world time. This relation of the time of mental events to that of physical processes is not merely one assumed for the purpose of experimentation,

---

[22] Alfred Schutz, *Collected Papers* in three volumes.
[23] See Richard Hönigswald, *Philosophie und Sprache* (Basel, 1937), especially pp. 53-66.

merely for the purposes of measurement. Rather, experimentation and measurement must assume this relation since it is only in terms of the significance of physical processes (as independent of any particular human experience and as the most basic fabric of mundane existence) that mental events have the sense of being in the time of the world. Objective world time is, after all, not the subjective experience of the individual but an ideal of measurement in terms of events independent of and open to all minds. The sense of the objectivity of mental events presupposes the structure of the physical world and the mind's location in it. That is, any recognition of the objective time of conscious events must be in terms of their most intimate physical correlates, since there is no independent objective mental time, only the objective time of physical events. This is not to deny that time is the form of inner intuition, or that any measurement of time presupposes (at least psychologically) the inner experience of time. Rather it is a question of the sense of objectivity and its presuppositions. Practically, this means one must select a standard of time in the world and then in terms of this standard all other events are timed, including those of the central nervous system. The central nervous system is then the standard for the time of conscious events since consciousness is located in the world through the body, specifically the central nervous system.[24]

In short, the body appears as one's location for perception and action. It is through identification with the body, and the space-time of the physical world which is potentially a common standard for all mundane minds, that mental life gains objective spatiotemporal location and significance. The objective space-time of a sensibly intuiting and acting mind is the space-time of its body which it presupposes as a moment of its significance. Thus, we are always within a spatiotemporal (physical) context and cannot escape it; we can move only to another similar and derivative context. How much and what of the world we perceive and can act upon is determined by our location spatiotemporally and historically. One does not have the perspective upon the world open to other minds whose bodies have a different location. One is always experiencing and active from a *here* and now, and other minds (if they exist) must always be experiencing and acting from a *there* which one can only appropriate by physically occupying that particular *there*. But the occupation of that position succeeds only in a later now. A mind is isolated from other minds by a difference of perspective in experience and action. One is consequently particularized in a potentially social space and time (i.e.,

---

[24] See *Husserliana,* vol. 4, *Ideen II,* #63.

history) that presupposes the spatial and temporal particularization of physical objects. One is physically located or embodied in the world.

6. *Community with the world*

Embodiment is primarily a relation between two strata of significance and only secondarily a relationship between two classes of entities.[25] That is, we wish to call attention to the categorial significance of our description. Specifically, we wish to point out that the relation which we are examining is a "founding-founded" structure between categories of appearance. On the one hand, the category of finite mind presupposes the category of body as a necessary moment of its significance. The body is the mind's location. On the other hand, the body possesses the significance of being a unified and complex mechanism of discrimination and operation apart from the significance of mind. This relation, this community, is asymmetrical. Mind presupposes the category of body as the foundation of its mundane significance. Body is a moment of the significance of mind, not mind a moment of the significance of body. Or otherwise put, the category of mind can encompass the significance of the body, but the category of the body cannot encompass the significance of mind. The category of mind is the richer category and part of its richness is that it can appropriate the category of body and be mind incarnate. The notion of ensouled matter reaches beyond the notion of body; the notion of embodied mind brings content within the notion of mind. This can be explicated through describing the mind's community with the world.

One is not in a world of pure geometrical forms, but rather of objects which are brute facts. Indeed, Kant indicated this: spatiotemporal objects are objects for receptivity, objects to be encountered as brute givens. The mode in which an object is present in reality, the mode of its existence, is even a more primitive note of significance than the particularizing notes of spatiality and temporality. It indicates most abstractly the domain of reality to which an object belongs. The presence of minds is consequently complex, for they are not just minds but embodied minds. They appear as the overcoming of the mere bruteness of physical objects,

---

[25] Husserl's analysis of the psychophysical reality of man in *Husserliana*, Vol. 4, *Ideen II*, #62-64, also reveals a founding-founded structure of significance. Moreover, a similar analysis, more explicitly in terms of strata of reality, is given by Nicolai Hartmann. See *Neue Wege der Ontologie*, especially Chapters IV and V (Stuttgart, 1949); also *Philosophie der Natur* (Berlin, 1950), and *Das Problem des geistigen Seins* (Berlin, 1933).

yet their presence is itself still brute in respect of their body. Thus, while physical objects are given as brute facts, minds in contrast are given as mere facts. That is, minds are given appresentationally,[26] as facts that add dimension and richness to the physical world through containing the brute presence of a body. Without this note of brute facticity the mind would fail to be part of the physical world. Simply put, it would not be one of its facts. This involves the notion of being in a world and the contention is that the most abstract note of significance possessed by the world is brute facticity. Things are just there. Or in more refined terms, things exist in contrast with ideas.[27] The very existence of ideas and things, their presence in reality, is distinct. For minds to be part of physical reality they must share in its mode of existence. They must be one of its facts. We will now explore this relation of community between the mind and the world which makes the mind not an idea, but an empirical reality.

To begin with, the world in which I am enmeshed is always to some degree cognitionally opaque. Unlike ideas which are translucent to thought and rational action, things give their significance adumbrationally, from a perspective, and as brute facts. Every experiential apprehension of an object is only a particular apprehension and not a total view. The full significance of physical objects is realized asymptotically. As a result, a physical object is always divisible *ad indefinitum;* it is never given completely and at once as a pure simple atom of impenetrability. In such a world knowledge and action must to some extent be estranged from their object. If in experience and action it were otherwise, if intuition and action were in immediate possession of their object, then the object would not be given as a thing (an to some extent opaque object, a brute fact); it would be rather given as an idea. Thus, integral to the very significance of a sensible spatiotemporal world is the eidetic requirement that all its loci and all their dynamical relations be physical. If not, they would be loci in an ideational world, a world of pure thought.

Consequently, a mind insofar as it is in the world is also a physical object. If a mind lacked such a significance, it would either be angelic or in relation to a physical world, it would be a mere idea. In order for mind

---

[26] The notion of the "appresentation" of other minds via their body is developed by Husserl in *Cartesianische Meditationen,* #49-54. For our purposes, appresentation indicates that minds are part of our experience, but only via their bodies. Minds appear as founded significances, presupposing the physical presence of the body.

[27] See Hegel's treatment of nature as "the idea in the form of otherness (*Anderssein*)" in the "Begriff der Natur" at the beginning of the *Naturphilosophie*. One can understand Hegel as isolating this first and most primordial significance integral to being an element of a physical world.

to have complex finite perception, it must include complex physical discrimination of objects. And in order for mind to be the instigator of complex actions within a physical world, it must include the initiation of complex physical operations. Otherwise, there would be intellectual intuition, not a finite apprehension of the brute givenness of a physical world, and miraculous alteration of the world, not the process of action. Sense perception and human action are, after all, the result of the interaction of subject and object. The senses appear as passive, not as creating their objects but as receiving physical presence. And the body in turn appears as active, not as merely the occasional cause of the creation of the design of the mind. To be in the world, to act and perceive in it as a finite mind, is to have the perspective of a mechanical organization of matter that reflects the unity, perceptive ability, and complexity of mind. Only as the mind is so present and located in the world does it possess, in the world, the discriminative ability and complexity requisite for action and perception. Physical analogues to mental occurrences are the basis or ground for mind's presence in the world. In short, location in the world is a physical involvement in the world.

This condition is but an instance of the general eidetic structure of being-in-a-world implied by Kant's third analogy.[28] It is a phenomenological foundation of the Presocratic principle of like being related to like. To be in a world is to be in community with its objects *as* an object of that world. The requirement by mind of a rather complex embodiment is an expression of the meaning of a type of community – that of perceiver and actor with perceived and acted upon physical objects. This community is given in one's experience of the body as his complex physical location in the world. Indeed, the organs of my body (as we noted above) distinguish themselves as various unique physical vehicles of my presence in the world. The complexity of this differentiation physically reflects the complexity of mental functions and operations. That is, these organs are the mind extended in space and time. As such these organs, in particular the nervous and muscular systems, must be adequate for the physical realization of experience and action. The body must be the locus of the physical reception, transformation, and retention of patterns in the world; the body must be the locus of the physical initiation of

---

[28] Wolfgang Cramer gives an interesting Kantian development of this notion in his *Grundlegung einer Theorie des Geistes* (Frankfurt/Main, 1965). He says, for example, "Indeed, not/S [= nature] continually acts upon S [= "that which is self-determining," for our purposes "mind"], because not S [= the organism] stands in a dynamical relation with not/S [= nature]." #49, p. 46. That is, the mind is in reciprocal relation with the world only via its body.

action. Only then does the body realize the significance "location of mind," and only then does it ground the mind's significance as mundane, as of this world. Without that necessary moment, the significance of finite mind ceases to be real and becomes ideal – it lacks a community with the physical world that could ground the significance of perceiver and agent in this world.

Moreover, this community leads to the ingression of physical significance into the significance of mind. A mind *has* its physical states such that it makes sense to say "I am here" when referring to where my body is. A person as an embodied mind is veridically described by descriptions of his body; for example: "*He* does not synthesize insulin properly; *he* has a biochemical or genetic defect." The significance of the body is invaginated into the concrete significance of the higher category, mind. The physical identity of the body is identified with the unity of consciousness. When one recognizes a person's living body, one recognizes that person. There is a continuity upwards from the objective identity of the body to the subjective identity of the mind. Yet on the other hand, the mind can be said to appropriate the unity of the body as an extension of its own unity. It is in and through the body that the mind senses and affects the world. In so possessing the body, the mind unites the body to the unity of its life. The body, itself a physical object devoid of mental significance, is made to coincide with the identity of a mind, the body becomes more than a body. It becomes the mind's incarnation.

Thus there is not only a presented difference of mind and body, but also their unity or identification in a single pole of identity. It is in the lived body that the identity pole of mind and of the physical body coincide in a founding-founded structure of significance. The interidentification of mind and body is termed an identity-in-difference, because it is a relation that has two disparate aspects. On the one hand, there is identification of mind with body, and body with mind. I am identified with my body as my existence in the world, and my body is identified with me in that I am its full significance (i.e., the purpose and unity of the body as an embodiment). On the other hand, the significances, mind and body, remain distinct and are not conflated. To stress this, the phrase "unity in difference" will be used interchangeably with "identity in difference." That is, the identity of mind and body is, as we have seen and shall see more clearly, not a formal logical identity, but the interidentification of categories of being.[29] Our description has thus led us to delineate a relation between domains of significance apparent in reality.

[29] But unlike Wolfgang Cramer, this thesis has to do with the interrelation of

7. *Final and efficient causality: further contrast and unity*

The identity in difference of mind and body is found in the relation between final and efficient causality. Final causality augments the meaning of merely physical occurrences so that they come within the domain of mental life. For example, a sensation causes my desire in the sense of effecting it. But the sensation does not merely *effect* the desire, it also has the sense of *soliciting* or *motivating* the desire. The desire's occurrence, instead of being external to the sense of the cause (i.e., the sensation), is a turning *towards* the sense or significance of the sensation, a desiring *of* the sensation. Moreover, the desideratum enters the domain of mind as part of a pattern of intendings. It is a reminder of X, it is recognized as a sensation to stop, or as novel and interesting, etc. One is caused to perceive the sensation because it is the term of a perceptual intending. It is a *sensum* because it is a *sentiendum*; it is a sense or significance that solicits sensuous perception; it is the goal of a perceptual intending. It can only be onesidedly, abstractly, termed an efficient causation. One can more properly say that the mind is solicited or motivated towards a significance, than merely that a state of mind is efficiently caused.

The purely efficient causality of physical objects lacks a dimension of solicitation or motivation. Impact does not *motivate* the body (qua physical object) to move, it *forces* it to move. The body is acted upon in a manner extrinsic to it; it responds to the impact without pleasure, aversion, volition, or thought, but according to a quantitative equation describing the general behavior of physical objects. This stratum of efficient causality, though, is the physical basis for final causality. Final causality appears only as a further enrichment of efficient causal processes; final causality is not a mode of the miraculous. Perception can solicit our interest in an object only because the object can make physical contact with us. That is, perception of an object and the object's motivation of our aversion or desire does not succeed via an immediate intuitive contact with the object. Rather the nature of a physical world condemns the mind to mediated relations (see Chapter II A6). The object repels us or draws us on via an efficient causal relation that underlies the possibility of any drive or aversion to particular objects. Yet in all of this the fabric of intentions remains distinct from that of efficient causality. Though an object may cause my perception of it, its stimulus is not the

significances of being, not realities as things. See *Grundlegung einer Theorie des Geistes.*

same as the object's solicitation of my attention. The latter occurs within mental life, and the former is a process recognized by mind as the basis of its connection with the world. Thus we have introduced a special use of the term "teleological causality" to designate the intentionality that overlays what are otherwise merely efficient causal relations. Such teleological causality is always a part of the fabric of a mind's life and is related to fabrics of efficient causality as the mind is to the body.

The contrast of solicitation with mechanical or physical causation is not blunted because some physical events are describable teleologically but without reference to intentionality. In such a domain appearances have the significance of being directed towards a goal, but without any reference to the presence of a mind. This would, for example, include physiological processes occurring for the self-maintenance of an organism, but without any imputation of feeling or other modes of mentality. Such end-directedness is a level of objective unity distinct from that of other domains of physical reality. It is a further level of comprehensibility integral in certain appearances. A quote from Jerry Fodor serves this point. "The remark 'a drive is not a neurological state' has the same logical status as the remark 'a valve lifter is not a camshaft.' That is, it expresses a necessary truth." [30]

For our purposes, which center upon man and his experience of the mind-body relation, we can avoid a separate treatment of teleology by examining it both as i) already integral to the life of mind, and ii) merely a part of nature and in contrast to mind. The first is legitimate since man is conscious and his life is intentionally teleological. Teleology in this case appears as a moment of mind. The second is also legitimate since teleology considered apart from mind is unaware of itself and can be considered as a special stratum of unity, but within the domain of physical reality.[31] We will later introduce a description of the human body as a

---

[30] "Explanation in Psychology," in *Philosophy in America,* ed. Max Black (London), p. 179.

[31] Though Hegel speaks of the animal organism as being a subject (*Werke,* Vol. 9, *Naturphilosophie,* #337) and of its having sensibility and sensations, his discussion of sensibility is primarily in terms of the body as a system of physical organs, as a stage of objective unity. He speaks for example of the life of the brain indicating a certain level of organization (*ibid.,* #354 Zusatz, p. 597). It would appear that sensation has quite a different meaning in Hegel's discussion of nature than in his discussion of mind. That is, it recurs in the *Philosophie des Geistes* as part of a new and higher unity of reality, and thus with a richer meaning. But whether Hegel is clear or not, we distinguish teleological processes that occur without reference to a mind from those that do. The former may even include drives, instincts, and needs, as well as irritability and sensibility as Hegel does. "Sensitivity" and "sensation" then would not refer to a feeling but to the fact that in being sensitive the organism is directed to its own self-preservation through the vicissitudes of its environment. As such,

universalizing mechanism. There the unity of the human body's physical processes sets it apart from non-organic bodies. It acts "as if" it were directed to its environment in order to master it, etc. Thus, in the case of the human body and other animal bodies, it would be possible to speak of non-intentional teleology. Instead, in order to focus on the experienced contrast of mind and body, we will term the causality of that unity "efficient causality" in that there is no awareness of the goal. It is efficient causality that appears "as if" it were aware, but without the note of being aware. Final causality we will stipulatively define as including awareness, albeit only a feeling, of the *telos*. The contrast of final and efficient causality is then in respect of inclusion or non-inclusion within the life of mind.

The contrast of final and efficient causality is not due to a reflective abstraction. Rather the contrast is itself presented in appearance, though such experiences are rarely associated with habitual acts. In habitual acts the body is fully significative of the mind's intent, fully enlivened by the mind's purposes. It is rather in the experience of the recalcitrance of the body, as a fabric of efficient causality centered about an objective identity pole, that the moment of physical significance becomes explicit. In such cases, one is confronted with the body's refusal to add to its stratum of efficient causality the fabric of one's final causality. For example, when one attempts to regain the use of a partially paralysed limb, he is confronted with a recalcitrant, but highly organized, material structure. The limb appears as a unity of efficient causality that resists mind's attempt to become more fully incarnate in it. This unity is not that of a mind and conflicts with one's attempt to control his body.[32]

Consequently, we can point out certain eidetic features of subject and object poles of appearance. Each has its own matrix congealing an object around a sense of identity. Each constitutes the fabric of significance of a category of appearance. On the one hand, efficient causality has no immediate place in the mind's fabric of intendings. Physical processes are as such anonymous and thus are not immediately united by a subject pole. The human body naturalistically considered is not one's body save legally. It lacks the personal significance, "*my* body." Its processes have as such no relation to a subject; a subject pole cannot immediately unite

---

being sensitive would mean "responding to a stimulus." (Dorland's *Medical Dictionary*, p. 1364.) It would be a note of unity transcending the merely mechanical, but it would not yet be the aware unity of mind.

[32] It is tempting to generalize and see in this relation the mind's relation to the world: a drive towards bringing the physical world within the domain of mental life and thus overcoming its alien character.

merely physical processes. They lack any reference to meanings, noemata; they are outside the life of mind. On the other hand, since mental events are intendings of noemata, they cannot be immediately united in an object pole. An object pole cannot unite intendings without taking on the sense of being a mind. But this is indeed what occurs in the case of the lived body: the body is a pole of identity for intendings and therefore becomes more than an impersonal body of matter. The anonymous status of a merely physical body is transformed into the sense "my body," "in the world"; it coincides with the unity of mind. This category leap is presented *de facto* in efficient causality's status as the substratum for final causality. The body considered in this respect is incarnate mind.

But only certain patterns of efficient causality are presented as embodying mind. These can be described as having a tendency towards more than mere physical significance. For example, the purely neurological processes associated with higher thought cannot possess their sense of being intricate, sensitive, discriminative, and patterned responses to patterns of matter and energy, without always carrying with themselves the possibility of being *understood* as the physical substratum of thought. Importantly, this does not beg the question. Rather, it calls to our attention the fact that these physical processes themselves are biased towards the stratum of mind, in which what is, experiences its being. The categorial leap is a move to a further richness that is appropriate for types of physical appearance that possess a spatial level of unity.[33] That is, certain patterns of physical processes have the sense of being onesidedly considered, if they are considered merely as physical processes.

One can descriptively explicate this tendency, since appearance affords variations of the characteristics of the phenomenon, "body," beyond which "body" ceases to be able to possess the significance: "embodiment of a mind." That is, finite minds cease to be located in this world when there is no physical embodiment of their intelligence and agency. When the body does not perform the physical processes of reaction to stimuli "as if" the body were intelligent and active, there is death. The body is in that case no longer appropriately the location of a mind. The criteria for the identification of the physical processes which are the embodiment of mind are isolable through specifying what sort of body is presupposed by a mind. Our description of the special importance of the nervous and

---

[33] This "tendency" is noted in Husserl's description of the psychophysical reality of man. "The natural-Object man is not a subject, a person, but every such Object accords with a person. So we can also say: every such object 'implies' a person, an I-subject." *Husserliana,* Vol. 4, *Ideen II,* #62, pp. 187-188.

musculoskeletal systems indicates that if a body is to incarnate an experiencing agent, then it must have physical processes of the same order of unity and complexity as the mental life it embodies. If the body does not possess a systematic complexity of the same order as the mind, then to that extent the mind is not in the physical world. This is seen in brain damage.

Conversely, when one observes an organization of matter that is able to respond systematically to complex patterns so as not only to preserve itself through environmental changes, but in addition to control its environment, etc., then one has found a system of efficient causes that could be described in final causal terms, including intelligent goals. In order to be the founding stratum of a mind these physical patterns of occurrences (in the case of human: neurophysiological processes) must i) be as complex and unified as a possible psychical description of them. That is, for example, the physical stratum of the "recognition of an object" must include refined patterns of the discrimination of objects accountable in terms of physical laws. Otherwise, mind is not given as objectively present in the world, but only metaphorically so. ii) A mind is a unity; components of mental activity cannot be treated as existing in isolation. Thus, one cannot impute a mind to a machine if all it can do is discriminate certain classes of objects, no matter how complex and refined that discrimination might be. iii) The patterns of the physical processes must be discrete and comprehensible enough to be recognized as united in and arising from a particular organized object. Otherwise, one would not be referring to a concrete presentation of mind; rather, the significance "mind" would be only abstractly present. iv) The physical parts of this complex object must be so interconnected that they present the sense of being a self-centered unified whole, especially noted as a whole in respect of its "intelligent behavior."

In short, the body must have a unity analogous to that of a mind. Otherwise final causality does not further constitute the sense of an object, but merely operates in a regulative or heuristic fashion in our search for order in nature. Or phenomenologically put, the object must appear such that the lack of a description in mental terms gives the description a sense of being essentially onesided. Further discussion and specification of these requirements must wait until later (Chapter V). Here it is sufficient to note that certain objects can be encountered that have the sense of being fully described only when they are described not merely as bodies but also as embodied minds.[34] The body must appear as the sign of the

[34] "The personal subject is the fulfillment of physical 'performances.' " Husserl,

mind and thus gesture beyond itself.[35] In this way the lived body is the presented unity of mind and body: it is presupposed by and betokens the mind. It is an epiphany of the founding-founded relation of mind and body.

8. Mind and body in neurophysiology

This excursion into the lifeworld of the natural scientist is a phenomenological perspective of the findings of neurophysiology and is an amplification of our previous description of the embodiment of consciousness in order to grasp in greater detail the impact of natural science. Or otherwise put, we will examine certain experiences which natural science affords. (See also Chapter I C 1bi concerning experience in the natural sciences.) Though these experiences are not a part of everyday life, they are a natural extension of it. They are experiences of a man when he adds to the domain of ordinary life the actions and experiences associated with the life of a natural scientist. Moreover, though the natural scientist operates within a theoretical attitude, it must be remembered that his experiences are not theoretical constructs (theoretical insights imposed upon straightforward experience to elucidate its efficient causal connections, though not themselves straightforwardly presented in experience; e.g., proposed mechanisms for ion transfer across cell membranes, storage and retrieval systems for memory, etc.) or inferred entities (never given directly in experience; e.g., potassium ions, sodium ions, etc.). Rather, they are present, directly accessible realities of the world. That they are novel and have new and but recently considered consequences should not give us pause when we remember that many of the primary realities of our world have been but newly introduced to us. For example, the existence of cells, bacteria, etc. – objects displayed to students in elementary biology classes – have but recently become a part of the domain of possible experience for most persons of our culture. In short, these realities are amenable to the same sort of phenomenological descriptions as those of everyday life.

Thus, what natural science reveals often presents itself in the same level of reality as most objects and experiences of everyday life; there are no sharp boundaries. For example, it is an experience of everyday life that cutting one's finger *causes* pain; that is, the cutting presents itself (to an

---

Husserliana, Vol. 4, *Ideen II*, Beilage XIV, p. 381. Also see *ibid.*, #62 and 63.
[35] See Hegel's remark concerning the body as the sign of the soul in *Werke*, Vol. 10, *Philosophie des Geistes*, #411.

adult) as the *cause* of the pain. Similarly, ethyl alcohol has the sense of being an intoxicant once one has had some acquaintance with it; it is a straightforward experience of ethyl alcohol as psychogogic. *Pari passu*, one can learn that specific drugs cause particular sorts of psychical states. Moreover, one can enlarge this experience by noticing (through electroencephalography, etc.) how the drugs concurrently change the operation of the brain. One can do this also in the case of ethyl alcohol and the cut finger. One can then become acquainted with the fact that the alcohol and cutting not only have effects upon one's psychological states, but also upon the physiological states of one's nervous system; one finds that not only do alcohol and cutting *cause* psychological states, but that they also *cause* physiological states. A next step (and one of the same degree) involves noticing that the occurrences of the psychic states are *correlated* with those of the physiological states. The correlation in fact becomes so striking that it takes on the sense of interdependency, as is the case in quite ordinary correlations such as that of craving for food and its physical manifestations: gastric hunger contractions, etc. The correlation of mental states and bodily states thus is encountered straightforwardly in the natural attitude. It is a part of everyday life amplified and enriched in detail by the experiences of the natural scientist.

As an example we will briefly review Dr. W. Penfield's record of a patient's responses to his stimulation of her cerebral cortex.[36] The experiments exhibit the dependency of sensory motor and psychical states upon the condition of particular areas of the cerebral cortex. They were performed by stimulating the brain via an electrode; the threshold for response was two to three volts of 0.2 milliamperes.

Stimulation of the motor and sensory cortex produced respectively bodily movements and sensations. Included in brackets are the anatomical designations of the cortical areas corresponding to the number of the stimulated area in Dr. Penfield's operations. "Stimulation at point 2 [postcentral gyrus] – sensation in thumb and index finger. She called it 'quivering,' 'tingling.'

3 [postcentral gyrus] – 'Same feeling on the left side of my tongue.'

7 [precentral gyrus] – Movement of the tongue.

4 [inferior portion of the postcentral gyrus] – 'Yes, a feeling at the back of my throat like nausea.' " [37]

---

[36] The material is excepted from a record of an operation performed by Dr. Penfield and presented in his Sherrington lecture, *The Excitable Cortex in Conscious Man* (Springfield, 1958).

[37] *Ibid.*, p. 28.

Stimulation of the interpretative cortex has been shown to produce psychical responses, that is, the stimulated person hears and sees events often associable with past events.

"11 [anterior portion of the superior temporal gyrus] – 'I heard something familiar, I do not know what it was.'

11 – (repeated without warning) – 'Yes, Sir, I think I heard a mother calling her little boy somewhere. It seemed to be something that happened years ago.' When asked to explain, she said, 'It was somebody in the neighbourhood where I live.' She added that it seemed that she herself 'was somewhere close enough to hear.' " [38]

This correlation of psychical states and physiological states is extensive and has been further illuminated since Penfield's first experiments. For example, psychical states caused by both external stimulation and cortical stimulation (as well as due to stimulation of structures within the brain) can be compared so as more fully to present the interrelationship of mind and body. Again a quotation from the report of an experiment is illustrative. "In addition, the stimulus to the skin or to ventral posterolateral (VPL) nucleus of thalamus is so located as to elicit a sensation within the same somatic area as that in which the sensation was subjectively felt when the recording site on somatosensory cortex was stimulated directly." [39] These events are not unique and the patient himself could have them repeated in future craniotomies as well as observing them by attending future operations. They are a part of the reality of the world. For the patient they are as available as ordinary experiences. For us, they signify the availability of a domain of experience in which the dependence of mind upon body (not the reduction of mind to body) becomes vivid.

Further, mental events also appear as the "cause" or the identifying marker for physiological processes. A most obvious example is voluntary movement. Every voluntary movement can have its precipitating cause (the action potential transmitted to the muscles involved) traced back to the brain and ultimately to the motor cortex. At that juncture it becomes impossible to isolate the volition as prior and thus temporarily distinguished as the "cause" of the physiological processes (vide Chapter II A5). The volition appears only as that through which we identify the physiological processes, since the muscular movements are the *terminus*

---

[38] *Ibid.*, p. 28.
[39] B. Libet, et al., "Responses of Human Somatosensory Cortex to Stimuli below Threshold for Conscious Sensation," in *Science* 158, (22 December 1967), p. 1597.

ad quem of the volition. The volition is our most direct acquaintance with the *terminus a quo* of the causal processes, though it can be replaced by direct electrical stimulus of the motor cortex, as Penfield and others have shown. Further, even mental states can be shown to "cause" marked neurophysiological reactions. An interesting example is provided by theta waves (waves recorded by electroencephalograph, and which have frequencies between four to seven cycles per second). ". . . they also occur during emotional stress in some adults, particularly during disappointment and frustration. They can often be brought out in the electroencephalogram of a frustrated person by allowing him to enjoy some very pleasant experience and then suddenly removing this element of pleasure; this causes approximately twenty seconds of theta waves." [40] The fact that physiological processes appear radically correlated with mental events (as "caused" by them) rounds out the picture of the interidentification of mind and body. One is presented with a radical interdependence of psychical and physiological states, a state of affairs that can be described in this manner: "In the conditioned reflex activity of the cerebral cortex the physiological process is accompanied by psychic phenomena – the subjective form of the physiological process." [41]

What natural science has done is to clarify further the unity of mind and body by showing their radical interidentification. The spatiotemporal, physical location of the psychical has been made more explicit. The physical presuppositions of our thinking, willing, perceiving, and acting are apparent in everyday experience (vide Chapter I A1). But in neurophysiology our mind is in addition presented as localized in certain portions of nervous tissue. We are presented with the geography of the physical locus from which we perceive and act. The direct phenomenological description of this geography is dependent upon experiments such as Dr. Penfield's. Such experiences further reveal the brain's significance as the differentiated localization of mind. Given this general framework, information from postmortem studies, etc., can be used to fill in the picture of the brain's geography with the result having more than the modality of an inferred or hypothetical structure required by a certain scientific theory.[42] This result is the appearance of a highly organized self-regulat-

---

[40] A. C. Guyton, *Textbook of Medical Physiology* (Philadelphia, 1961), p. 807.
[41] E. K. Sepp, "Localization of Functions of the Cerebral Cortex," *The Central Nervous System and Human Behavior* (Maryland, 1959), p. 445.
[42] "Over the past 50 years, several investigators have mapped the cortex according to variations in structural characteristics. Although the original maps indicated some 20 areas, more recent work has subdivided many of these into a grand total of 200 or more zones. Experimental and pathologic research has demonstrated specific

ing system of matter which has the significance of being a spatially differentiated physical incorporation of mind. One is shown the geography of the embodiment of mind.

By revealing the central nervous system's intricate embodiment of mind, science presents us with an astounding relation between domains of reality. When one finds organized matter fulfilling the criteria of proper neurophysiological functions, and responding to its environment in a manner characteristic of human bodies with higher cerebral function, then one *eo ipso* has a human body that is presented with the sense of embodying a mind. This coincides with the fact that in seeing another's living body we have the sense of seeing him, not merely a physical body. This thesis is not psychological, but rather ontological. It is as if such an organized body dialectically implied a mind, as if it fulfilled itself naturally in a higher level of reality.

That is, far from revealing the reduction of mind to body, neurophysiology strengthens our previous conclusions concerning their founding-founded relation. Their experimental interidentification coincides with the lived unity of mind and body. The mind enters objective reality (objective spatiotemporal relations) only via its body. And the body enters the world of intentions of relations only via the mind of which it is a moment. Further, the terms identified remain distinct, as they do in ordinary experience. A mind is of necessity not a mere body, nor is a mere body a mind, even when one term is made the identifying term and the other the term identified. Onesided statements do not erode this. If one abstractly focuses only on the physical aspect of this unity, one approaches a reduction of the psychical to the physiological. Consider, for example, this statement of Sir C. S. Sherrington to Dr. Penfield, " 'It must be fun,' he said, 'to speak to the preparation and have it answer you.' " [43] If one takes the next step and asserts that the mind is reducible to the physiological processes, one overlooks the fact that the psychical and physiological are presented as co-realities. It is for this reason that such a step causes an intellectual shock. The sense of the mind and that of the body are categorially distinct; they cannot be collapsed without a paradox. An identity pole of subjective events cannot simply be an identity pole of objective processes without losing its presented significance. As Wittgenstein's ordinary man remarks, "*This* is supposed to be produced by a process in the brain! – as it were clutching my forehead." [44] But the contrast does

functional differences among a large number of these zones." E. L. House and Ben Pansky, *A Functional Approach to Neuroanatomy* (New York, 1960), pp. 418-419.
[43] Penfield, p. 5.
[44] *Philosophical Investigations,* tr. G. E. M. Anscombe (Oxford, 1963), #412.

not deny the fact that mental events and physiological processes appear so intimately connected that the unity suggests that each is just the other's other side.

The contribution of neurophysiology has been the enrichment of our notion of the identity in difference of mind and body so that the body's role in this relation is specified more clearly. It reveals the fantastic complexity of the physical substratum of mind as well as the particular relation between the mind and this complex organization of matter. Neurophysiology displays the intricacy of embodiment. It is not as if we did not or could not know of this before. That is, in other circumstances we would have been pressed to posit such a complexity (see Chapter II A6). Neurophysiology has allowed us to confront it firsthand.

B. A PHENOMENOLOGICAL OUTLINE OF AN ONTOLOGY

In the course of our investigation, we have descriptively isolated the eidetic features of the relation of mind and body. In particular we have become acquainted with their unity in difference. As descriptions of the eidetic features of being, they are also propositions in an ontology (see Chapter I C1). Body and mind are after all categories of being-for-us, and their founding-founded relation is a basic structure of reality. It is the relation between fundamental levels of significance; it is a categorial relation. We can summarize the results of this descriptive or phenomenological ontology and thus outline a categorial account of mind and body.

a) There is a relation of opposition or distinction between the body, which is a unified matrix of efficient causality, and the mind, which is a unified matrix of final causality or intentionality. Each constitutes the unity and significance of a domain of appearance and each is distinct. They are categories.

b) Besides the distance between these strata of significance, there is a union. Mind and body are articulated in a founding-founded structure in which the body is presupposed as a necessary moment of the mind's significance and in which the body has a tendency to be understood as a moment of mind.

i) The mind presupposes the body for the significance of being an experiencing agent in a spatiotemporal sensible world. The body is the ground upon which the mind's mundane significance is based. That is, the category of mind without the moment of physical reality is not the category of mind that stands in finite perceptive and active relations with a physical world. To be a mind in the world is to be located physically and in physical community with the world. Moreover, this involves a physical complexity commensurate with the complexity of the mind's perception and action. Perception and action are in the world only insofar as they are also physically present: embodied. This requirement is emphasized by neurophysiology which reveals the geography of embodied mind and the radical dependence of the mind upon its physiological substratum.

ii) But this dependence has another and interesting side. That category of bodies which possesses the internal unity and complexity requisite for intricate discrimination of and operation upon objects tends to betoken the category of mind. The category of the complete mechanism for discrimination and operation has the unique sense of being encompassable as the physical moment of mind's existence. There is nothing extra-physical in the body to describe but this categorial bias: the body can be apprehended as the mind incarnate. This disposition or tendency is the expression of the possibility of presupposition. Moreover, this asymmetrical disposition-presupposition structure of the founding-founded relation of the categories of mind and body further specifies their relation. The body is the lower presupposed category which is disposed to being more fully understood in terms of the encompassing founded category of mind.

iii) Thus, the categorial contrast between the physical body that is lived, ensouled, and the mind that lives the body, that besouls it, does not precipitate a dualism, for they are united as elements of a founding-founded relation that effects their unity as embodied mind. Their identity poles contrast but coincide. The reciprocal structure of presupposition and disposition constitutes the categorial relation of mind and body. That is, there is presented *de facto* a relation of identity in difference between the categories of mind and body: in their distinct status each presupposes an identification with the other. The phenomenal equivalence of mind incarnate and besouled matter is the coincidence of the subject and object poles of appearance in the appearance, lived body.

c) This structure of mind and body is evident even in the particularization and presentation of the mind. The body is the ground of the mind's presence and particularity in the world.

i) The mind's particularization presupposes the spatiotemporal particularization of the body.

ii) The presence of the mind involves the co-givenness of the body. The mind is presented through the body, in that in being given the body, one is also given the mind.

d) In fine, by describing the relation of mind and body as an identity in difference we recognize their categorial significance. Neither category can be satisfactorily treated in isolation, nor can they be conflated. Rather, the presented structure of the significance of these categories is a founding-founded structure of meaning constituted from the asymmetrical references each category makes to the other. Integral to this asymmetry is the appearance of the founding category's significance encompassed in the new dimension of significance provided by the founded category. The category of body is encompassed by the category of mind, though distinguishable from it. Mind and body are distinct inseparables; they are categories distinguishable in thought, though mundane mind is as such inseparable from its embodiment. The encompassment of both these dimensions, mental and physical, is requisite for an account true to the appearances.

CHAPTER III

ALTERNATIVE ACCOUNTS

Chapter II has provided an ontology of mind-body. Phenomenology is the explication of being-for-us (vide Chapter I C 1c). Though the ontology's claim to veracity lies in mirroring the essential structure of reality, it is not complete by itself; nor is its necessity fully evident. In order to secure our ontology, we will examine the possibility of alternatives. Indeed part of this ontology's significance is the exclusion of alternatives. By excluding both dualism and monism it establishes itself as an ontology of a special kind; it describes an identity in difference. But an alternative which it does not exclude defines its incompleteness, and will take us to the question of the dialectic. Thus, in order to develop our account we turn to others.

A. CONFLICTING ONTOLOGIES

Ontology has traditionally tried to delineate the categories of mind and body and their relation. As the relation is conceived, so also are the two categories, since their significance is determined within a founding-founded structure. And the reverse also holds, since the relation arises out of these categories. An ontology which onesidedly isolates either the unity or the difference of mind and body, or denies the reality of either mind or body, commits a fallacy of onesidedness. This fallacy is the assumption that a fundamental diversity of categories is incompatible with a concurrent fundamental union, or the reverse. The first form of this fallacy is the supposition that the fundamental relationship between mind and body is their contrariety, so that all unity is secondary. The second form is the supposition that if a unity underlies the relation of mind and body, then one of them must be reducible to the other.

The first supposition formulates the relation of mind and body in a paradoxical dichotomy. The body as the embodiment of the mind and the mind as the enlivener of the body are less than easily comprehensible if there is no unity rising out of the significance of the categories. In that case, they are merely contingently or externally related. This ontological stance is the outline of a dualism: the fundamental relation of mind and body is found in their independence, and their intrinsic unity is not recognized. But, instead, one can consider only the mental or physical strata of this relation, trying to deny or reduce the other. This is the foundation of a monism: the fundamental relation of mind and body is their unity or identity, so that there is no essential difference between them; what appears to be mental (or physical) is held to be really but a special manifestation of the physical (or mental). This forces one to describe the sense of the mind-body relation and its properties ultimately either in physical or mental terms – the one chosen determines the nature of the monism.

As a result of these onesided analyses, the history of the investigation of the mind-body relation has been the history of a problem. Those more sensitive to the distinction between mind and body constructed dualisms with the consequence that the unity of mind and body could not be satisfactorily explained. Those more sensitive to the unity of mind and body, or to the reality of either mind or body, constructed reductive monisms, and failed to explain the difference between mind and body.

We will examine both sides of this dilemma and diagnose the nature of their failure. Each represents a misinterpretation of the relation and thus offers a crucial albeit negative lesson. Each provides an illuminating contrast with the account provided by our phenomenology. By understanding the failure of its alternative, the virtue of Chapter II's ontology will become more apparent.

1. Dualism

The principle of dichotomy has produced dualist theories of several varieties: interactionist theories, espoused by, among others, Rene Descartes; parallelist theories which include those of Leibniz, Malebranche, and Spinoza (according to some interpretations); epiphenomenalist theories which distinguish mind and matter, allowing matter to act upon mind but not the reverse (as Feigl described it, mental states are nomological danglers from physiological states), etc. All these positions so stress the contrast between mind and body that no way (within our own terms) remains to explain the relation. Spinoza recognized Descartes' difficulties in this regard. "What does he understand, I ask, by the union of the mind and body? What clear and distinct conception does he have of thought most intimately united with a certain small portion of matter?" [1]

This difficulty is expressed in the question: how can mind and matter interact? One cannot answer if the mind and body are ultimately distinct. Indeed, by the hypotheses of a dualism, the realm of the mental is inexplicable in terms of the physical and vice versa; there is nothing in the physical domain with which the mental is of its nature related, and conversely. An essential discontinuity prevents one from building a bridge of understanding between these domains. Such an ontological geography shows two contiguous but totally different terrains of being. The reason for the unity of these diverse domains is not apparent from them, and their affinity is in the nature of a problem.

Of course, not all dualistic accounts are pure, and therefore the degree of difficulty varies. Indeed, in the case of Descartes, there are at least two themes of analysis. One favors a strong distinction between mind and body, making any union between them a fact immediately and wholly dependent upon God's ordinance of the world rather than upon the significance of the categories.[2] Consider, for example, Descartes' assertion "I am therefore strictly only a thinking thing, which is a mind, or soul, or intellect, or reason . . ." [3] Unless carefully qualified, such an analysis of man implies that man's relation to his body is a mere contingent surd fact. "It is more worthy of admiration that a mind is to be found in each human body, than that there is to be found no mind in any of the animals." [4] Such an interpretation precludes anything more than a paramechanical model of the union of mind and body in which, for example, the mind moves the pineal gland, regulating the animal spirits in order to move the body.[5]

This paramechanical model fails to be anything more than an *ad hoc* postulate, for there are no clear grounds upon which to justify the possibility of action across the ontological gulf between mind and body. As

---

[1] B. de Spinoza, *Opera Quotquot Reperta Sunt,* Vol. 1, *Ethices,* Part V (The Hague, 1914), p. 247.
[2] See Rene Descartes, *Oeuvres,* Vol. 7, *Meditationes de Prima Philosophia* VI (Paris, 1904), pp. 80-90.
[3] *Ibid.,* p. 27. It must be remembered that the possibility of prescinding the notion of mind from that of body does not mean that this can be accomplished without paying the price of onesidedness or great abstractness. Vide Chapter II.
[4] Descartes, *Oeuvres,* Vol. 5, *Correspondance,* Descartes a Morus, 5 February 1649, p. 278.
[5] Descartes, *Oeuvres,* Vol. 6, *Dioptrique;* Vol. 11, *Traité de l'Homme* et *Les Passions de l'Ame,* especially Part 1, articles 34 and 35. Also note the anatomical plates in this volume which represent the pineal gland's physiological functions and emphasize the explicit location by Descartes of the mind's operation in the body.

Spinoza pointed out, Descartes "distinguished the mind from the body to such an extent that he was unable to assign a single cause to this union or to the mind. Rather, he had to recur to the cause of the entire universe, that is to God." [6]

In asserting that both mind and body are fully understandable in isolation from each other, Descartes has precluded anything but a fortuitous union; he has onesidedly prescinded from the presented union of mind and body. Consequently, he cannot understand their union in terms of their categorial significances. These categories, separated as distinct ideas, lose the dynamic relations that they possess in a founding-founded structure of significance. Instead of being able to explicate the significance that the body has a mind, he is forced to account for the possibility of the interaction of two distinct substances. The unity of mind and body which is experienced in the founding-founded structure of appearance is classified as an unphilosophical assessment of reality.[7] As a consequence, the notion of the body as mine becomes problematic, for it is in actuality (vide Chapter II A) grounded in the mind's presupposition of the body as a necessary moment.

Descartes' position is the outcome of a choice between onesided alternatives. Descartes admitted this in a fashion paradigmatic of what Hegel termed the abstract onesidedness of the Understanding (which distinguishes categories while ignoring their interrelations). "It does not appear to me that it is possible that the human intellect can at the same time very distinctly conceive of the distinction of mind and body and their union. That is the case, for to conceive of them as one single thing and at the same time to conceive of them as two is a contradiction." [8] But as our phenomenological observation has shown, this is indeed the "contradiction" that appearance presents us: the identity in difference of mind and body. Descartes has gone astray by artificially over-separating the categories of mind and body so that there is no account of their union.

Alongside his classical dualist analysis, Descartes showed a sensitive appreciation of the union of mind-body, absent in Malebranche. Descartes not only recognized interaction between mind and body, but asserted that we experience the fact of our union with body – a very non-occasionalist almost phenomenological observation. "Nature besides teaches me by these senses of pain, hunger, thirst, etc., that I am not present in my

---

[6] Spinoza, *Opera*, Vol. 1. *Ethices*, p. 247.
[7] Descartes, *Oeuvres*, Vol. 3, *Correspondance*, Descartes a Elisabeth, 28 June 1643, pp. 691-692.
[8] *Ibid.*, p. 693.

body as a pilot in a ship, but that I am so intimately conjoined with my body that it is as if I were thoroughly mingled in it. This is to such an extent that I compose one thing with my body." [9] Thus Descartes claimed to operate with a concept of the union of mind and body[10] such that at least in a certain fashion the mind is not merely accidentally conjoined to the body. He asserted, for example, "for the human body to be united to a soul is not accidental to the body, but of its very nature. The body has all the dispositions requisite to receiving a soul and without which it is not truly a human body. It would be impossible, short of a miracle, for the soul not to be united to the body. Furthermore, it is not accidental to the soul that it is joined to a body; rather, this is only accidental to the soul after death when it is separated from the body ... And on that account, as I said a short time ago, the union is accidental *to a certain extent* but it is not *absolutely* accidental." [11] Unfortunately, Descartes did not illumine the ground of the possibility of this union. Instead, we are invited to recognize a condition inexplicable in terms of Descartes' usual contrast of mind with the body as a mere automaton.[12] The notion of experienced unity of mind and body is not employed by Descartes because he considered it an adventitious idea, not the result of philosophical investigation. "It is by using only ordinary life and ordinary considerations, abstaining from meditation, and instead attending to the things which excite the imagination, that we can learn to understand the union of mind and body." [13] Thus, the union is understood only at the expense of philosophical thought, which is as much as to say that the union is surd. It is philosophically unjustified.

Surprisingly, at times Descartes conceived of automata not as mere machines but as sensitive organisms, though devoid of higher thought. "I would wish that it be noted that I speak of thought, not of life or feeling. I do not for instance deny life to the animals, alleging that it consists solely in the heat from the heart. Nor do I deny feeling (*sensum*) to the animal insofar as that feeling depends upon corporeal organs." [14] This

---

[9] Descartes, *Oeuvres*, Vol. 7, *Meditationes* VI, p. 81.
[10] See Descartes, *Oeuvres*, Vol. 3, *Correspondance*, Descartes a Elisabeth, 21 May 1643, p. 665.
[11] Descartes, *Oeuvres*, Vol. 3, *Correspondance*, Descartes a Regius, mid-December 1641, pp. 460-461.
[12] See Descartes, *Oeuvres*, Vol. 11, *Traité de l'Homme* et *Les Passions de l'Ame*, Part 1, article 6, etc., for a discussion of man's body as an automaton.
[13] Descartes, *Oeuvres*, Vol. 3, *Correspondance*, Descartes a Elisabeth, 28 June 1643, p. 692.
[14] Descartes, *Oeuvres*, Vol. 5, *Correspondance*, Descartes a Morus, 5 February 1649, pp. 278-279.

is quite provocative, since a thinking being is also a "res ... sentiens"; [15] thus a trichotomy is engendered among thinking reality, corporal sentient life, and purely physical reality.

Since ours is not an examination of Descartes *per se*, we need not weave these conflicting threads of insights into a more coherent pattern. Instead, we take them as manifestations of the tensions within a dualist analysis. Within such a framework mind and body no longer portray the founding-founded relation presented in experience. As a consequence, the separated and opposed elements, mind and body, are alienated from their experienced union. These difficulties, caused by the imposition of a metaphysical schema upon experience, invite a further strengthening of the dualism by denying the union or interaction of mind and body. This involves denying the element of appearance which is a pole of tension with the dualism: the presented union of mind and body. It obviates the difficulty of explaining how such disparate categories (as classical dualisms conceive mind and body) could interact. Instead, the union is found in another principle, outside of either mind or body. Still, the union appears surd or arbitrary, for an examination of appearance does not reveal a third principle but a founding-founded structure between the categories body and mind. (Vide Chapter II.)

Nicolas Malebranche illustrates such an analysis. He claimed a strict separation of mind and body and he denied that any union arose out of their respective categorial significances. Instead, the union was attributed to God. "Our soul is not at all united to our body as is maintained by common opinion. The soul is united immediately and directly only to God." [16] In short, he constructed a parallelist account of the relation of mind and body. The natural or physical causes are in harmony with our mental life through the will of God and therefore have the status of being only occasional causes. "For it is evident that there is no necessary connection at all between the volition we have, for example, to move our arm, and the movement of our arm. But *natural* causes are by no means the true causes; they are nothing but *occasional* causes which have no efficacy save by the power and efficacy of the will of God." [17] The union of mind and body is a fiction of common opinion.

Since the will of God is eternal [18] there is only a short step from Male-

---

[15] Descartes, *Oeuvres*, Vol. 7, *Meditationes de Prima Philosophia*, p. 28.

[16] N. Malebranche, *Oeuvres Completes*, Vol. 12, *Entretiens sur la Métaphysique et sur la Religion* VII, #15 (Paris, 1965), p. 168.

[17] Malebranche, *Oeuvres*, Vol. 2, *Recherche de la Verité* VI, II, III, p. 315.

[18] See Malebranche, *Oeuvres*, Vol. 12, *Entretiens sur la Métaphysique et sur la Religion* VII, #9.

branche's occasionalism to the preestablished harmony of Leibniz. "The soul follows the laws peculiar to it, as the body also follows its own. They agree in virtue of the preestablished harmony between all substances, for they are all representations of one and the same universe." [19] In such a framework, the relation of mind and body is explained via the will of God and not the categories which appearance presents. This constitutes an explanation of the categories of experience, not in terms of their logic, but in terms of a transcendent principle. It neglects the presented structure of experience and moves away from determining the character of the world we live in and experience. Not only does this entail a tension between the metaphysical account and being, but it also invokes an explanatory principle that is not itself suggested by the specific problem at hand. God is not immediately a part of the significance of the founding-founded structure of mind and body. The introduction of God as an explanatory principle has the sense of being *ad hoc* or even mysterious; *voluntas Dei ignorantiae asylum est*.

With the dichotomy of mind and body, the notion of unity is lost sight of in an attempt to conceive of the possibility of interaction between such diverse substances. The focus of attention does not fall upon the significance of my body as me in the world. It falls instead upon the question of the nature of the causality involved in uniting mind and body. This results in neglecting categorial analysis of presented being for speculation concerning the possibility of mechanical action between the soul and the body. And such categorial analysis is not overlooked by chance, since the initial interpretation of mind-body does not start from presented significance, but from a priori theoretical construction of being that precludes an encompassing account.

For this reason radical dualists such as Leibniz are constrained to adopt and defend an explanatory principle that is *ad hoc* and surd in relation to the explanandum mind-body. The virtue of such a principle is not that it is given in the fabric of being, which it is not, but that it does a good job of elucidating the mind-body problem which dualism engenders. Thus, Leibniz asserts in defence of his account, "I confess that the phrase *preestablished harmony* is an artificial term. But it is not a term that does not explain anything, since it is made out very intelligently." [20] Thus, rather than having an exhibition of the structure of the significance, mind-body, one has an attempted solution to difficulties associated

---

[19] G. W. Leibniz, *Opera Philosophica, Monadologie* (Meisenheim, 1959), #78, p. 711.

[20] *Opera*, fifth letter of Leibniz to Clarke, #90, p. 774.

with the "mind-body" problem. But these solutions arise out of the initial dualist construction of the mind-body relations, not out of its presented significance. The onesided perspective of the dualist thus leads him away from being and towards the interminable difficulties of solving a mind-body problem engendered by neglect of an eidetic feature of being.

In fine, dualisms violate the principle of encompassment (see Chapter I C lc) and thus fail to give a satisfactory account of the being of mind and body. 1) They overlook an element of experience that we find to be eidetic in the very significance of its appearance: the founding-founded unity of mind-body. 2) They consequently give an account of the categories, mind and body, that i) portrays them as far more independent and distinct than the examination of appearance reveals, and ii) forces a unity upon them that is not natural to them, but surd (i.e., its necessity is not comprehensible in terms of that to which it is applied). Reason is left dissatisfied.

2. *Monism*

A monism is often an endeavor to escape the difficulties of a dualism by denying the very foundation upon which dualism is erected. It denies that the most fundamental relation of mind and body is their difference. Its postulate of reduction asserts that, since ultimate dichotomies are incomprehensible, it follows that in any dichotomy one term must be reducible to the other. In the history of the analysis of the mind-body relationship the reduction has been applied both ways. There have been monistic analyses that have, like Hobbes, reduced mind to matter, and those that have, like Berkeley, reduced matter to mind. Though distinct, these analyses share a difficulty. To account for the mind-body relation a monism must account for the apparent significance of the mind-body distinction; this entails accounting for the distinction itself as well as the term that has been reduced.

To accomplish the first a monism must either show that what the dualists recognize as a difference in kind is only a difference in degree, or they must reject the distinction itself as an illusion. But it is eidetic in experience that minds and bodies appear qualitatively distinct. (See Chapter II.) Therefore the monist must in some way account for the apparent difference in kind in terms of degree by denying the reality of one of the terms. This leads one to the second part of the problem: one must show how mind has the properties of matter or matter the properties of mind. This reformulates the mind-body relation on a new level as two aspects of one "substance." In order to establish a monism one must show that one aspect is *really* the reality and the other not *really real* but *merely* an aspect. For example, if one collapses the domain of mind into that of body, one must credit the body with mental properties. But then one can ask for an account of the body's having mental characteristics, and recover the question of their relation. Consequently, a monism must engage in some process of reduction in order to cancel out the reality of one of the poles of a dualism. The reduction must show that the significant marks of one stratum of reality are totally accountable in terms of the ingredients of "another" stratum of reality and thus that a dualism is a redundancy.

Historically, a number of forces have favored such a reduction, especially a reduction of mental reality to physical reality with the consequent establishment of physical monism. Principal among these forces have been the remarkable advances and explanatory powers of the physical sciences. They have in particular inspired a program to reduce mental states to states of the nervous system. This has led to considering man primarily in terms of his nervous system and the emergence of what Wilfred Sellars terms the concept of a "core" person. "Imagine a person who has been defleshed and deboned, but whose nervous system is alive and in functioning order. Imagine its sensory nerves hooked up with an electronic system which enables it to communicate. Without expanding on this familiar science fiction, let me simply suggest that we think of what we ordinarily call a person as a nervous system clothed in flesh and bones. In view of what we know, it makes perfectly good sense to introduce the term 'core person' for the empirical nervous system and to introduce a way of talking according to which raw feels and, for that matter, thought are in the first instances states of 'core persons' and only derivatively of the clothed person." [21]

The crucial question then is the manner in which mental states are to belong to neurological processes, for if the belonging is understood as an identity, then they are in fact reduced away. The question is whether the core person is completely described when he is described neurophysiologically. U. T. Place and J. J. C. Smart, among others, assert that in fact the true description of the core person is a neurophysiological one. Smart maintains that "sensations are nothing over and above brain processes ... When I say that a sensation is a brain process or that lightning is an electrical discharge, I am using 'is' in the sense of strict iden-

---

[21] Wilfred Sellars, "The Identity Approach to the Mind-Body Problem," in *Review of Metaphysics* 18, (March 1965), pp. 441-442.

tity." [22] He is seeking to avoid the complexity and nomological difficulties of a dualism [23] by unifying the study of man in a fashion that leads to the greatest explanation and prediction.

But the interest is not merely instrumental. Rather, it is ontological, for these theorists are not satisfied with merely asserting that natural science is essentially the mode of quantitative explanation and prediction (and that therefore insofar as one is interested in a quantitative study of man, he must reduce human reality to physical facts). On the contrary, the human person is held to be a physical entity; the intent is ontological. "So by a combination of the mechanistic and contextualistic categories, I believe the neural-identity theory can hold up. If it can, it breaks through the wall of dualism between mind and matter that has pestered mechanistic naturalists from the time of Descartes." [24] The concern is to give ontological foundation to an expedient procedure in natural scientific investigation, the reduction of mental events to physical correlates for purposes of certain studies.

Monism's difficulty is the presented ontological reality of both mind and body. A reduction of either becomes more difficult the more one attends to the categorial distinctiveness of mind and body. As a consequence, the dualist position often reappears in a disguised form. Herbert Feigl's account of psychical states instances this. Though he asserts "according to the identity thesis the directly experienced qualia and configurations are the realities-in-themselves that are denoted by the neurophysiological descriptions," [25] he still maintains that not only is there a stratum of merely physical appearance, but also part of this stratum lacks at least an accessible reality-in-itself. "Instead of conceiving of two realms or two concomitant types of events, we have only one reality which is represented in two different conceptual systems – on the one hand, that of physics and, on the other hand, where applicable (in my opinion only to an extremely small part of the world) that of phenomenological psychology." [26]

The dualism persists in the form of a bipartite conceptual system presented in two distinct strata of being-for-us. What Feigl has done is to stipulate that one of the categorial strata is uniquely being-as-it-is-itself.

---

[22] "Sensations and Brain Processes" in *The Philosophy of Mind*, ed. V. C. Chappell (New Jersey, 1962), p. 163.
[23] *Ibid.*, p. 162.
[24] Stephen C. Pepper, "A Neural Identity Theory of Mind," in *Dimensions of Mind*, ed. Sidney Hook (New York, 1960), p. 60.
[25] "The 'Mental' and the 'Physical'," in *Minnesota Studies in the Philosophy of Science*, Vol. 2, ed. Feigl and Scriven (Minneapolis, 1963), p. 457.
[26] "Mind-Body, *Not* a Pseudoproblem," in *Dimensions of Mind*, p. 41.

This fails to solve the difficulty because i) the selection of experienced qualia as reality-in-itself is founded on the expedience of avoiding either a Cartesian dualism or Spinozistic double aspect theory, not on the eidetic structure of appearance. It is merely an *ad hoc* assumption brought forth to solve an ontological riddle, not a feature of being discovered in phenomenological or categorial investigation.

ii) Even if one accepted that experienced qualia are realities in themselves, one is still left with the difficulty of accounting categorially for the totally distinct stratum of significance constituted by physical reality. Nor will Feigl's extension of the meaning of "denotation" solve the difficulty, for he has merely stipulated that the relation of mind and body is such that "the data of experience *are* the reality which a very narrow class of neurophysiological concepts denotes." [27] Obstinately, it remains *only* a stipulation because the significance of physical appearance, even that of neurological processes, does not denote mental reality, but does in fact present physical reality (see Chapter II). Feigl attempts thus to define the identity in difference of mind and body into a simple identity. But the contrasting significance of both categories of reality remains. Indeed, Wilfred Sellars recognizes this while forwarding an account similar to Feigl's. "In other words, I accept the identity theory in its *weak* form according to which raw feels or sense impressions are states of core persons, for, as I see it, the logical space of raw feels will reappear *but unreduced* in a theoretical framework adequate to the job of explaining what core persons can do." [28]

In short, a monism is not effected, but rather the dualism of mind and body is restructured. To show that either mind or body is an act, aspect, function, relation or way of appearing of the other, is not sufficient. The opposition or contrast between mind and body reappears as the contrast between the significance of the accident and the significance of the substance in which it adheres. All the problems of dualism can reassert themselves and challenge one to show how such a substance could have such a disparate property. Therefore, this dilemma appears: one either has a dualism at the level of the mental and physical "properties," or one must deny a dimension of reality. But, if one denies the reality of mind, one denies that there are lived experiences and contradicts the facts of the matter. Or again, if one denies all reality to matter, one again denies a significance that certain appearances possess (see Chapter II).

Some monists have taken the heroic step of denying absolutely the ex-

---

[27] "The 'Mental' and the 'Physical'," p. 453.
[28] "The Identity Approach to the Mind-Body Problem," p. 449.

istence of one pole of the mind-body relation, thereby hoping to preclude a recurrence of a dualism at any level. It is worth again noting that in physical monisms this is usually motivated by a desire to give an absolute ontological status to the expedient move in natural science of ignoring certain domains of reality. Since the move is solely to aid quantitative, merely physicalistic study, the expedient requires only that a domain be in fact isolable. It does not require the non-existence of other domains of reality. Indeed, the ontological status of mind remains untouched by the fact that it is not *really* real within the context of certain laws of natural science, for the naturalistic attitude is precisely the act of prescinding from other than physical reality. The naturalistic attitude merely reveals reality that has been truncated for certain purposes. (See Chapter II.) On the other hand, the demand for an absolute reduction of all reality to purely physical reality is a dogmatic program for absolute knowledge within the naturalistic stratum of significance.

B. A. Farrell exemplifies this. "What they would like to have is the assurance that they – as physiologists and neurologists – can in principle give a *complete* account of what happens when we think, recognize things, remember, and see things; and that they are safe in ignoring the mind and all intervening mental events whatever." [29] This absolute, though onesided, grasp of reality is to be assured by a certain cultural revolution initiated by natural science. "Contemporary science, in short, does not seem to require the notion of 'experience' and is getting to the brink of rejecting it, in effect, as 'unreal' or 'non-existent.' If the relevant sciences go on developing in this direction, and if Western societies assimilate their work, then it is quite possible that the notion of 'experience' will be generally discarded as delusive. If and when this happens, our present philosophical difficulties about it will disappear." [30]

One cannot, though, exclude categories of reality, unless one is no longer concerned with ontology, but with the instrumental requirements of a particular scientific discipline. That is, as a neurophysiologist one may have to bracket the further significance of certain physiological events in order not to be distracted from treating the material at hand neurophysiologically. Whether that is possible or really helpful is another question, and one surely open to dispute. But it is not the case that science as such can legitimately deny a dimension of reality. Empirical science, our systematic empirical knowledge of the world about us, would then be asserting a general monistic axiom concerning the structure of being.

[29] "Experience," in *The Philosophy of Mind*, p. 24.
[30] *Ibid.*, pp. 44-45.

Indeed, to label a category of experience ontologically unreal is a category mistake, unless one means to assert that the category is not in fact an eidetic feature of a particular type of world. In that case, the dispute can only be resolved by phenomenology. Otherwise, one is asserting that what is in fact essential for a type of appearance is *not* essential.

Yet the controversy can be allayed when, for example, one recognizes that the assertion "insofar as a sensation statement is a report of something, that something is in fact a brain process" [31] can usefully only refer to considerations within the enterprise of natural science. In that case, Smart would be asserting that ultimately the most inclusive account of reality productive of both explanation and prediction will be developed by the natural sciences and that such an account will, through natural laws, give a verifiable explanation of human behavior. This of course would in no way affect the ontological category "mind," which remains a general and essential characteristic of the world. Nor does it preclude a not so inclusive science, one whose only task would be to describe and explain the further reality of the restricted class of entities: embodied minds (e.g., psychology). But the only relevant argument against the category mind must return us to experience, where the category was purportedly discovered.

The program of a radical physical monism thus involves an attempt to banish a dimension of reality, not simply because it is not amenable to natural scientific treatment, but because this recalcitrance is an ontological embarrassment for monism. A physical monism seems to offer, at least to many, the simplicity and uniformity of explanation that has attended the progress of natural science; but the possibility of a dualism compromises this ambition. Consequently, an attempt is made to dismiss as mischievous the recognition of a uniquely mental stratum of significance, and bring the dispute within physically tangible terms at any cost. If one accepts the monism/dualism disjunction, then this may appear to be the only choice unless one is going to dismiss the entire question as senseless, as have Wittgenstein and some of his followers.[32]

Finally, if the monist program is completely to exclude the possibility of the recurrence of a dualism, it must include some forms of "double think." B. A. Farrell again serves as an excellent example. He enjoins us to hasten the day when one will no longer be tempted to think of *mental* reality. "Some positive recommendations... Get rid of the nuisance

---

[31] Smart, "Sensations and Brain Processes," p. 163.
[32] See Blue and Brown Books, p. 8; it should be added that Wittgenstein's rich comments do not wholly support the move to ignore the problem.

words like 'sensation,' 'experience,' and so on, by defining them provisionally by means of concepts like: stimulus patterns, a discrimination by an organism, a readiness to discriminate, a discrimination of a discrimination ... They *can* get rid of mental events and experience. But they get rid of them in a queer way – by realizing it is just foolish to suppose that there are, or are not, such things." [33] In short, radical physical monism is an act of ontological faith that succeeds by ignoring a dimension of reality. Consequently, it fails because i) there remains a dimension of reality untouched and unaccounted for.[34] Moreover, ii) the dimension that is described remains onesided in its isolation and, because of its disposition toward indicating the possibility of a fuller account of reality (see Chapter II B8), it invites the challenge of a dualism to recur. Importantly, these conclusions hold also against a mentalist monism. It too must fly in the face of appearance by denying a category of sense that appearance presents and indeed which the category of mind presupposes (see Chapter II).

In the case of psychical monism, one must be careful to distinguish i) the attempt to exhibit the grounding of all significance in mind and ii) the attempt to *reduce* all significance to mental significance. The first is not a move to a monism but to a clarification of the meaning of significance in general (see Chapter I C2). The second, though, is an attempt to deny that there is a real distinction between mind and body. Psychical monism alleges that the "otherness" of body is merely an illusion. It overlooks the fact that though all notes of being are significant only for a mind, still some of these predicates constitute a physical object.

The distinction of mind and body is an eidetic feature of appearance, albeit appearance is only phenomenal, only for mind. That is, the body i) has the sense of having an objectivity transcendent to any particular mind, ii) has the sense of being an object for consciousness but not itself a consciousness. iii) The body is not pure idea immediately given to mind. Unlike a pure idea, it is a complex of significances particularized spatio-temporally and thus capable of being had from an indefinite number of

---

[33] "Experience," pp. 46-47.

[34] A recent article in the *International Journal of Psychiatry* correctly diagnoses the quandary of the physical monist. "In conclusion, the mind-body or neuropsychiatric split remains a focus of discomfort for by far the majority of psychiatrists who are concerned with theory. A monistic conception would appeal more to our sense of 'theoretical elegance'; it would satisfy our need for consistency and would appear more congruous with prevailing scientific positions. However, we cannot, in a form of obstinate psychological denial, bluntly state that the problem does not exist. ..." Silvano Arieti, "The Present Status of Psychiatric Theory," *International Journal of Psychiatry*, Vol. 8, (September 1969), p. 621.

spatiotemporal perspectives. At best, it is an alienated idea, presented not immediately in thought but mediately through adumbrative sense perceptions externalized in space and time. iv) In any case, there is a presented contrast of mind and body as identity poles of two different types of significance (see Chapter II).

As to whether certain idealists, as for example Berkeley, actually wished to deny the presented "reality" of the body is a historical question that falls outside of this investigation. It is worth noting, though, that Berkeley's analysis calls our attention to the absolute status of mind. "... What is meant, by *the absolute existence of sensible objects in themselves, or without the mind.* To me it is evident those words mark out either a direct contradiction, or else nothing at all." [35] This does not entail a psychical monism; his analysis of causality may, though, in that it compromises the presented otherness of physical objects by obscuring their status as identity poles of physical processes.

Further, the idealism or panpsychism of such philosophers as Charles Hartshorne is exempt from the charge of psychical monism in that the "physical" reality of aggregates of "atomic life" is recognized. The contrast of different levels of reality is maintained, though all levels are asserted to be in some sense "mental." "Thus if mind can be ... extended and also inextended ... if mind can be on both sides of every genuine contrast, in the sense in which reality can be so (even unrealities are at least mistaken but really occurrent ideas, fancies or what not), then there is no need to assume some additional principle." [36] At a certain macrolevel the body may thus legitimately be recognized as an "other" to mind; as the presentation of extended mind, the body's "physical" recalcitrance before one's intentions is properly a part of reality. Consequently, a onesided categorial view of reality is not espoused. In addition, such panpsychist theory is not so much concerned with the analysis of ontological categories as with the *physis* of things.

A true psychical monism (i.e., one that denies the significance of the category of physical reality) is probably found in Hindu thought. It enjoins an ultimate conflation of all categories of significance in a supreme category of mind, Brahman. "The attributes which distinguish phenomenal objects from Brahman are the result of maya and are therefore illusory. They are really one with Brahman, which, on account of maya,

---

[35] *The Works of George Berkeley Bishop of Cloyne*, Vol. II, *The Principles of Human Knowledge* (London, 1964), #24, p. 51.

[36] C. Hartshorne, "The Case for Idealism," in *The Philosophical Forum*, 1, (Fall 1968), p. 10.

appears as the universe and created beings." [37] The difficulty with this position lies in its denial of the reality of the category of body by dismissing it as *merely* an illusion. It excludes a category of being rather than attempting either further to analyze instances of it or further to comprehend it in higher categories. For example, without committing oneself to a onesided view of reality, one could have asserted i) that physical objects are *really* conglomerates of psyches, or ii) that all reality is only fully comprehensible in terms of its apprehension by a Supreme Mind. The fallacy of onesidedness, as in the cases of physical monism and dualism, resides in the failure to recognize a feature of being.

Such accounts, as mere denials of a feature of being, run counter to the presented fabric of experience, as well as failing to possess a possible categorial richness that would embrace both the mental and physical significance of being. They are intrinsically abstract, for they fail to be true to the concrete fullness of being (i.e., a being that includes both mental and physical reality in its ultimate significance). Thus the onesidedness of such an account contrasts with the concreteness (i.e., the faithfulness to being and its richness) of an encompassing account.

Interestingly, the failure of the program of monism returns one to a dualist dichotomy, since these onesided investigations presupposed for themselves a mutually exclusive character. The result is thus a dialectical oscillation between monism and dualism: the onesided account of mind-body as a dualism proves so unsatisfactory that one is pressed to deny a dimension of reality; but when this denial proves unfeasible, one is again returned to the problems of dualism. Thus one must either abandon the problem as a nuisance or decide that Descartes was wrong in maintaining that one could not at one time consider mind and body as both distinct and unified.[38] If the question is to be answered and the dialectic to be fulfilled, then we must unite these contraries in a notion of identity in difference and encompass all sides of the reality of mind-body.

3. Double aspect theories

Accounts which make mind and body aspects, functions, or different ways of knowing the same thing are in principle better conceived than either monisms or dualisms. Double aspect theories endeavor to preserve the distinction between mind and body without essentially dis-

[37] Swami Nikhilananda, *The Upanishads,* Vol. 1 (New York, 1949), p. 167.
[38] See *Oeuvres,* Vol. 3, *Correspondance,* Descartes a Elisabeth, 28 June 1643, pp. 691-692.

uniting them. But they imply a substance of which both mind and body are aspects or attributes. Such an account of the relation of mind and body requires an explanation of how mind and body are related through this third thing. Moreover, to avoid appealing to a mysterious noumenon, the third thing must be characterized. If it has other than mental or physical properties, they are unknowable, since mental and physical properties exhaust the set for concrete objects. If it is describable in terms of mental and/or physical properties, the problem of specifying their relation has been merely restated, save for the *assertion* of an identity in difference. In short, double aspect theories have the virtue of recognizing that the relation of mind and body is an identity in difference. But they merely assert such an identity. They do not start from the givenness of the founding-founded structure of the mind-body relation nor derive the identity from the sense of the categories involved. The identity appears rather as a fortuitous *ad hoc* assumption.

Spinoza presents an interesting example of a double aspect theory that attempts to free itself of the onus of accounting for a "third thing." Though he distinguishes mind and body as attributes of one substance, he presents the substance in which they adher not as an unknown third thing, but as appresented through any of its different attributes. That is, he strives to equate each attribute with the essence of substance and thus in effect to make substance the ground of unity between mind and body. Spinoza says in *Ethics*, Part 1, Definition IV, "I understand by attribute that which the intellect perceives of substance as constituting its essence." [39] He implies that each attribute "denotes" (to employ Feigl's usage) the same essence; this interpretation of Spinoza is fortified by his statement that each attribute "expresses eternal and infinite essence." [40] Thus, according to Spinoza, whether one knows via thought or extension, one knows the same thing; they are merely two different ways of knowing the same thing. ". . . Substance as thinking and substance as extended are one and the same substance, which is now comprehended under this attribute and now under that." [41] According to Spinoza, substance is not unknown, but known and understood through its attributes. Even if the attributes are infinite in number (Part. 1. Def. VII), and man is only acquainted with two of them, man (on Spinoza's account) still knows the essence of substance. That this is so, is also implied in Spinoza's assertion

---

[39] *Opera*, Vol. 1, *Ethices*, p. 37.
[40] *Ibid.*, p. 37.
[41] *Ibid.*, part 2, Prop. VII, p. 78.

that men have an idea of God which is adequate.[42] Spinoza's substance is not a noumenon.

If substance is not to be a mysterious *thing* that is both known and infinitely not known, then it must be understood to serve a certain categorial function. One can, for purposes of the mind-body relation, interpret Spinoza's doctrine of substance as a principle of unity. Spinoza is asserting that the two orders of reality are neither essentially independent of each other nor conflated. "Prop. VII. The order and connection of ideas is the same as the order and connection of things." [43] Similarly, "the mind and the body are one and the same individual which at one time is conceived under the attribute of thought and at another under the attribute of extension . . ." [44] Spinoza's double aspect theory thus exhibits the unity of being in the notion of substance and the distinct reality of mind and body in the distinct significance of the attributes of thought and extension. For once both the unity and distinction of mind and body are equally real (at least on the phenomenal level, that is, for men).

4. The fallacy of reification and the categorial turn

But Spinoza's theory, as other double aspect theories, fails to proffer a satisfactory account in that it is phrased in a "thing" language so that substance persistently suggests not a categorial relation between attributes but inherence in a *res substratum*. This is reflected in Feigl's criticism of Spinoza. "I reject the (Spinozistic) double aspect theory because it involves the assumption of an unknown, if not unknowable, neutral ('third') substance or reality-in-itself of which the mental (sentience) and the physical (appearance, properties, structure, etc.) are complementary aspects. If the neutral third is conceived as unknown, then it can be excluded by the principle of parsimony which is an essential ingredient of the normal hypothetico-deductive method of theory construction. If it is defined as *in principle* unknowable, then it must be repudiated as factually meaningless on even the most liberally interpreted empiricist criterion of significance." [45]

Thus, though Spinoza makes one of the most successful attempts to understand the unity of mind and body, he is still hampered by his frame of reference. It construes the unity of being in a fashion suggesting that

---

[42] See *Ethices*, part 2, Prop. LXVII.
[43] *Ibid.*, part 2, Prop. VII, p. 77.
[44] *Ibid.*, part 2, Prop. XXI, p. 95.
[45] "The 'Mental' and the 'Physical'," pp. 449-450.

substance is a "third thing" of which thought and extension are attributes. He fails to recognize the difficulty as a categorial one. He does not appreciate the integration of thought and extension as the integration of two modes or dimensions of significance. This failure and the failure of those who conceive of mind and body as distinct and possibly competitive types of reality can be termed the *fallacy of reification*. They have seen the issue of the "mind-body problem" as the question of whether there are two *res*, mind and body – and if so, of how they are related. Whimsically, one could term this the *substantial fallacy*: to overlook the notional nature of the relation of mind and body, to deal with the relation as the interaction of substances or of different aspects of substance.

In contrast, the transcendental turn is a turn away from things and towards their significance to thought. As such, phenomenology and especially eidetic phenomenology presuppose a transcendental turn – or at least what might be termed a categorial turn. Phenomenology reveals the categories of appearance and treats them not as things interconnected causally but as dimensions of significance interrelated in virtue of their meaning. It is thus that the mind-body relation is described as a founding-founded relation between categories of being. Phenomenology interprets mind-body and accounts for its unity by revealing the structure of its meaning. The shift is from substance to significance. Within this new frame of reference the old quandaries are without significance. There can be no question of reducing categorially distinct levels of significance one to the other. The richer level possesses notes of significance not conceiveable within the lower level. This distinction of the higher level assures its integrity though it presupposes the lower level as a moment of its own structure. The lower level on the other hand is likewise distinct in that it constitutes an independent though impoverished level of significance that can be contrasted with the higher level. They are recognized in their unity and distinction as existing in an identity in difference. Again, this refers to a certain connection of significance, not to a variety of interaction. From this perspective the causality of mind and body is neither interactive nor parallel but is appreciated as a categorial complex phenomenon that can be grasped in terms of categorial levels.

A distinction must be made here between adequate and encompassing explanation. The force of the argument has been that an account of man that excludes his mental reality fails to be encompassing. It leaves out a dimension of his being. This can be taken to imply that neurophysiological explanations, for example, are restricted from a particular domain of entities – i.e., minds. Under such a paradigm, the field of beings would be partitioned between psychology and physics. This is far from the meaning of the foregoing. Rather, since all reality with which we are confronted is embodied in this world, it is indeed physical (see Chapter II A). Thus physical science must include in its purview all entities. What it fails to include is a dimension of the reality of certain beings, which are not only physical but also mental. Encompassment refers to embracing dimensions of reality, while adequacy refers to embracing instances of reality. In regard to mundane reality, then, the term "physical" extends to and includes all beings, while the term "mental" extends to only a very *restricted* class of entities.

Adequacy and encompassment are two different axes of explanation. With regard to adequacy one is concerned that there be no gaps, no class of beings not considered; with regard to encompassment one is concerned that a dimension of reality has not been onesidedly excluded. The first relates to things and the second to categories. The contrast between the two is one of the contrasts between empirical science and ontology. Again the "mind-body problem" of monism has been a failure to recognize this contrast and thus the paralogistic attempt to seek adequacy of explanation where instead encompassment should have been sought. And the "mind-body problem" of dualism has been the failure to comprehend the interrelation of mind and body, since it was assumed that each belonged to a restricted domain of reality and that there was no underlying inclusive domain. The categorial turn allows one to recognize the depth of reality without trying to project the distinct dimensions of the depth upon one level of reality and thus, so to speak, force them to vie for adequacy. Mind and body are strata of reality, not mutual restricting, competing, and interacting domains of entities.

Having completed this chapter's survey of dualism and monism, one is then returned to the conclusions of the second chapter, now seen with a consideration of past failures. The phenomenological elucidation of the identity in difference of mind-body represents a move to a new perspective, which extricates this investigation, *eo ipso*, from the difficulties of the past. One has made a turn towards seeing the central questions of ontology as categorial questions. One has begun a categorial analysis of the significance of being. This turn from things to thought is part of the modern Copernican revolution in ontology. Through it the recalcitrance of the thing is overcome. That is, by no longer trying to divine the general rules for the interaction of things, but by instead attempting to delineate the general categories of the significance of things, one brings being into a perspective from which thought is competent to achieve its goals. Being is approached as it is for mind. The substantial fallacy is the attempt to discern eidetic rules for the comportment of things considered as independent interacting entities. One is then engaged in the strange task of promulgating logical rules for existing entities – one is engaged in the paradox of trying to mix thoughts and things. The categorial turn is the move towards seeing reality from a viewpoint that can ground things in thought and thus obviate any need for a mixture. One tries to describe the conceptual structure ingredient in existence – to phenomenologically explicate being for us. It is this change in approach that preempts the old questions of monism and dualism. The old questions cease to have a basis because the incongruities that incited them are now understood from a perspective that calls for answers neither dualist nor monist. The questions no longer concern two things, but rather two domains of significance.

One may perhaps be startled by this and wonder if we are not trying to hide difficulties by slight of word – by some involved variety of intellectual parlor trick. But one should remember that the mind-body problem has its natural origin out of a difficulty in understanding the relation of mind-body. The phenomenological answer provides the strains of thought ingredient in the reality of mind-body and thus answers the quandary – it tells us how to think of mind-body, it explicates its unity. Further, nothing is whisked away and no dimension of the quandary is banished as imaginary. The perennial question of the interaction of mind and body is answered, but as it turns out, in terms of an analysis of the complexity of action that always presupposes a physical substrate. The result is not an answer in terms of types of causality but in terms of dimensions of causality. The shift from things to significance does not entail a loss of any questions or possible answers but their translation into a new idiom.

This, though, does not make the move a merely verbal one. One is concerned rather with gaining a perspective from which an analysis becomes possible. The new idiom represents the concepts which are now able to be used given the new perspective. The move is not merely linguistic but is a move towards making it possible for one to successfully give the grammar of the language of being – ontology. All thought is linguistic in the extended sense of being conceptual. The categorial turn is more than a merely conceptual move in that it allows a successful descriptive analysis of being free of the difficulties that have accrued to other approaches.

A further move, though, is possible and necessary if a stronger justification of this endeavor and of the relations it reveals is to be given. This move has been termed the transcendental turn, the justification of the categories of being (and *eo ipso* of categorial analysis) in terms of reason's goal of encompassing explanation. This will be the topic of Section B below. Here it is enough if one notices that the categorial turn implicit in phenomenology and the rejection of reification represent a radical break from previous analyses of mind-body. Chapter V will show in greater detail how the old questions concerning the interaction of mind and body can be answered in a new fashion. At this juncture it suffices that another way of conceiving the unity of mind-body promises a departure from previous paradoxes. That in itself justifies pursuing this approach further.

In summary, we break free of the old monist-dualist dispute by leaving behind a reifying mode of thought through what can be termed a categorial turn. We shift our attention to conceptual consanguinities that may have nothing to do with picturable reifiable relations, but which are indeed the unifying structures of reality. Categorial analysis, insofar as it is successful, must broaden our notion of real relations. Its central assertion is that the unifying structures of reality are not physical but conceptual, that the relations between diverse, categorially distinct domains of appearance arise out of notional necessity, not out of material forces. This project of understanding the relation of mind and body through the relation of types of significances brings us then to examining the categories of mind and body in their relation to thought. By the terms "fallacy of reification," "categorial turn," "transcendental turn" we hope to indicate the difference between a phenomenological analysis and the traditional metaphysical analyses. We wish as well to indicate a direction of further and novel investigation (the transcendental turn).

The transcendental turn takes us beyond the phenomenological requirement of ontological encompassment. Ontological encompassment is necessary but not sufficient for a complete account, since mere ontological encompassment ignores the ground or reason why such a categorial structure is in fact discovered as necessary. It exhibits the categories but does not give a categorial account of the categories. We must take our investigation yet another step if we are to go beyond an unreflective acceptance of the eidetic structure of being. This further step is one of comprehending the reason for the *de facto* structure of being in terms of the goals of thought.

This step is implicit in some of the accounts that we have examined. Indeed, appealing to God as the ultimate guarantor of veracity (i.e.,

Descartes), as the ultimate foundation of harmony between levels of being (i.e., Leibniz), as the unity of Being itself (i.e., Spinoza), or as the Supreme Agent responsible for the coherence of appearance (i.e., Berkeley), can be seen as an embryonic appeal to a unique viewpoint taken to be the ground of significance. But the turn is neither actualized nor its possibility recognized, because these principles of the unity of the categories remain external to their significance. The principles of identity, contradiction, and sufficient reason, or the rules established by God for the succession of ideas are not the result of the grounding of the categories of appearance in thought. They are abstract in not being a notional unity that grows out of the significance of the categories. For example, they are not concrete as are Kant's constitutive categories or his principles of reason [46] which function as the necessary unity of the material significance of being. But in any case they are part of accounts that are intrinsically onesided. We are thus left where we began – with our phenomenology of mind-body (Chapter II). Unlike its alternatives, this account has been true to the identity in difference of mind-body. We have explicated reality and seen that our account follows a general canon of ontology (i.e., the principle of encompassment). But the *quid juris* is still to be answered. We must see if we can certify the truth of our ontology and ground its structure by passing from ontology to transcendental ontology.

B. TRANSCENDENTAL REQUIREMENTS

Eidetic phenomenology exhibits categories as well as the relations integral to their appearance. We think here of the founding-founded relation or identity in difference of mind and body. Eidetic phenomenology is an immediate confrontation with these categories and their relation. It provides a sketch of the *de facto* structure of being. Reason, though, can ask for more. It hopes to justify the relations of categories in terms of its own needs. Reason seeks unity and explanation. It can attempt to unify categories through their relation to the goal of explanation. It will be our contention that such a project is not overly optimistic. Rather it merely requires that we make explicit a notional unity that must be implicit in the categorial structure of appearance. To introduce us to this project we will briefly examine the "I think" and its bearing on the categories and their relations.[47]

---

[46] See *Critique of Pure Reason,* A 663=B 691.
[47] This transition to transcendental ontology presupposes and to some extent develops our account in Chapter I C2. The references to the "I think" are obviously

All elements of appearance, and most especially the categorial elements of appearance, are united in thought since experience is always uniteable in an "I think." The "I think" is the objective focal point of experience; it is an epiphany of the unity of thought. Only in virtue of being uniteable in thought do the categories of appearance and their relations receive the posited rationality requisite for their significance as the fabric of an objective world transcending the idiosyncratic status of any finite subject. If the categories of appearance are not related to thought and its unity, then they are mysterious modes of the idiosyncratic. In contrast the objective world is eminently rational; even its irrationality can be given a place. Its truth stands waiting for *any* man. This anonymity refers back to the "I think" as a posited theoretical stance through which the objectivity of the world is understood via the posited singularity of truth.[48]

But the unity of the categories cannot exist in the absence of an affinity. One is dealing after all with the unity of notions in thought, the womb of all concepts. Thought is the milieu in which categories and categorial relations are recognized – they are the structures of appearance apprehended in thought. The essential relation to thought guarantees the possibility of explicating the relation of the categories of appearance in terms of the goals that reason sets for thought. Or otherwise put, the unitability of experience in an "I think" is the implicit grounding of being-for-us in thought. If one elucidates this grounding in terms of the goals of reason, then one should be able to order categories as a series of stages of further explanation bound together via a founding-founded structure which is itself further illuminated in terms of the goal of explanation. This need not be seen as a radical departure from our previous phenomenology. Rather one could say that we have now begun to make the structures of self-consciousness thematic. We have turned to the rational unities that thought gives to the categories and categorial relations of appearance. We will focus upon our consciousness of the categories of appearance which are presented to consciousness. This, though, undoubtedly involves a change of levels.

As a consequence one could say that there is a two-story structure of categories. The first story contains the categories and their relations as revealed through a phenomenological description. As simply a description, explication, or indication of the structure of appearance it is termed

Kantian in inspiration, though not Kantian in consequence. They rather develop the implications of the relation of experience and thought in a direction different from that by Kant in the First *Critique*.

[48] The "I think" is thus a presentation of the theoretical stance or objective viewpoint of thought outlined in Chapter I C2.

de facto. The second story contains the rationale of the categories. In order to give a *de jure* account of the founding-founded structure between mind and body we will show *why* the founding category requires the founded category in order to complete the explanation of the material of the founding category. The relation of categories will be understood as a stepwise movement towards richer categorial comprehension. This stepwise movement is justified because it orders the categories in terms of reason's nisus towards a greater embrace of reality. It articulates them as various stages on the way to comprehension of the richness of being, which would be the ideal of reason.

This involves a dialectical metamorphosis of appearance, not the introduction of alien elements. The object of consciousness is further conceived in order to satisfy the demands of reason. For example, the full notional unity of the categories of mind and body is integral to their phenomenal founding-founded structure; but this is not apparent in a mere description of appearance. Rather, this appearance must be further and explicitly considered in terms of the goal of explanation. In general, this can be described as a dialectic of subject and object. The phenomenological object, the phenomenological category, fails to satisfy the demands of the subject and this requires a richer notion of the object: lower levels of reality must be seen as requiring, indeed implying, higher levels. This in turn causes us to develop further the notion of the subject. The phenomenological observer becomes one who also explains the phenomenological object in terms of reason and thus transforms the object. But again this does not involve going behind appearance but developing or making explicit what is necessarily a part of appearance. What appearance was all along in itself becomes presented to us, the phenomenological observer of appearance. The attempt to explicate the manifest structures of appearance leads naturally to the attempt to comprehend the reason for these structures. The move to ground the categories in thought is the modest proposal of attempting to understand the categories in terms of the demands of thought, the medium in which the categories appear.

This takes us beyond our investigation in Chapter II. Though we there developed an ontology that satisfied the requirement of encompassment, we did not ground the structures of appearance in thought. It is because of this further possibility that that account is incomplete. We must now show why, in terms of the logic of those categories, they form a unity for thought.

Thus the notion of an explanation's *completeness* is added to the notions of *encompassment* and *adequacy*. It concerns understanding categories in terms of the goals of reason and within our terminology is the goal of a transcendental account. The second, encompassment, is the more general goal of ontology, not to overlook a category of reality. The last is the goal of empirical science, namely, to be able to account for (explain and predict) the comportment of all classes of entities (i.e., not to be restricted from investigating any group of beings). Each signals a new dimension of explanation. Encompassment is richer than adequacy in that it embraces the latter while indicating a further direction for explanation. Similarly, the goal of completeness develops the goal of encompassment. The notion of completeness represents the appreciation that categories of being that encompass more reality, and which are thus less onesided and more self-explanatory, represent a step towards reason's goal of fully realizing an explanation of being. This third axis of explanation is the logical dynamism that is not recognized through the ontological principle of encompassment; it expresses the direction of the dialectic. It, so to speak, indicates the eros of reason. That is, completeness is the ontological requirement of encompassment viewed in terms of the *telos* of self-sufficient explanation. Completeness is encompassment not only with a view towards embracing the *de facto* dimensions of appearances, but transformed or *completed* by an orientation towards a goal of explanation which lies beyond a mere description of the appearances themselves.

CHAPTER IV

A TRANSCENDENTAL ONTOLOGICAL ACCOUNT

> All realities are related to one another, *affinitas*.
> The universe of realities (*omnitudo*) as a whole (*universum*) is to be posited as constituted in itself as a system of ideas.
> Immanuel Kant, *Opus Postumum I*, p. 107.

How can it be that a man is a body and a mind? This question can also be put this way: how are we to understand the relation of like categories of mind and body? It is this categorial interest that defines the nature of our investigation and brings us to the project of a transcendental account. What is asked here is not a question of empirical facts or even of the eidetic structure of the experience of a body and mind together. We begin with what is given. Rather, the question concerns the being of minds and bodies explicitly considered as grounded in thought.

The three previous chapters have developed canons for this investigation: i) the investigation cannot reduce mind to body or the reverse: these categories can neither be absolutely equated nor absolutely separated; ii) therefore one must show how mind and body can be both identical and different in the sense of being related as the same being but yet as distinct categories (while also avoiding the concept of an unknown underlying substance); and iii) such an account cannot be merely an exhibition of the eidetic nature of the founding-founded structure of the relation, but must also justify the structure and its elements by reference to the demands of reason, to the grounding of being in thought. In summary, the question of this investigation is: "what is the dialectic of meaning implicit in the identity in difference of mind and body?"

A. A DIALECTICAL RELATION

1. Introduction to terminology

In this brief subsection we will introduce terms employed during the course of this chapter. They will be explained in the context of their use, so that when they are encountered in subsequent sections their full import can be more easily recognized. The discussion accordingly is a preliminary abstract of the dialectic.

In the affinity of mind and body one is struck with their contrast. The opposition is inherent, for a lack of opposition obtains only in a onesided perspective. The opposition is itself given in appearance and does not yield to a reduction. Yet a reduction expresses the radical interdependence of mind and body and that is also given in appearance. The interdependence is inherently part of the relation and is likewise absent only in a onesided perspective. Consequently, our understanding vacillates between grasping the unity of mind and body, and grasping the distinct realities of mind and body. The vacillation of the understanding has its objective correlate in the dual significance of the mind-body relation: mind as embodied mind versus body as besouled matter. There is here both a unity and a contrast of mind and body, with a dual assertion of both realities.

This vacillation expressed by the alternation of dualistic and monistic accounts in the history of philosophy is presented in life when the body, due to some shattering incident or scientific attitude, is experienced as an alien thing. The fundamental status of mind-body, their intimate union, can become sundered so that one alternately has the sense of being a mind over against a body, or of being a mind thoroughly dependent on a body.[1] This oscillating dialectic will be termed an incomplete dialectic, since each alternative denies the other without establishing an encompassing level of explanation.[2]

Appearance presents the mind as presupposing the body and therefore having its sense founded on it. This relation has been termed one of "presupposition"; that which is necessarily presupposed as an element of a type of appearance is termed an "essential moment" of that type or category. Conversely, the intact human body appears to imply the human mind (see Chapter II A7) and founds the sense of the embodied mind. This relation has been termed a "founding-founded" or a "grounding-grounded" structure or relationship. The founding sense or ground is termed the "lower" level category or type of appearance and the founded or grounded sense is termed the "higher" level category or type of appearance. In that the higher level category preserves the lower level category as a moment (i.e., "sublates" it) while endowing it with a higher level significance (e.g., the body becomes the lived body or the enlivened body), the higher level category is said to "fulfill" the sense of the lower category

---

[1] R. M. Zaner has described this oscillation in "The Radical Reality of the Human Body," *Humanitas* II, (Spring, 1966).

[2] It thus resembles the false dialectic which Hegel describes in *Werke,* Vol. 8, *System der Philosophie 1: die Logik,* #81 (2).

(or be its "higher truth"). This founding-founded relation between two such distinct and contrasting categories will be termed a "dialectical" relation. Unlike the phenomenological founding-founded relation, it has the additional note of progression. The higher category is required in order to fulfill the lower; it is not merely discovered as the fulfillment of the lower category. The dialectic is the dynamics of the development of ever richer strata of significance.

The term "sublation" indicates the categorial encompassment of the lower category within the higher.[3] The lower category is not only a moment of the higher category, it is also uniquely explained by the higher category. The term "explanation" refers to a further enrichment of meaning, as when a process is no longer merely considered as the physical process of perception, but *as* perception (i.e., it is recognized as the psychical occurrence in which a physical process becomes "aware" of itself and in that sense explains or reveals its meaning to itself). The higher category is thus "higher" in that it supplies a further dimension of significance in which the "lower" category can be more fully significant. The higher category develops and completes the lower.

In the case of mind's relation to its physical body, this higher category can be said to make the significance of the body exist "for the body" (or explicitly), not merely "in it" (or implicitly). That is, the embodied mind is a body that exists "in and for itself," for it literally "realizes" its own significance in thought. Prescinded from this self-realization, the body is "alienated" from its higher truth. Its significance as a complex discriminating machine is merely "in itself," and for us the observers, but not yet "for itself." The body does not "realize" its notion save as embodied mind.[4]

It should be noted that our use of these terms is specialized. "In itself existence," "existence for an other," "existence for itself," and "in and for itself existence" are general categories of the mediation or determination of successively richer domains of significance. The categorial move from body to mind is but a particular, though crucial, example of the determination or making explicit of immediate, implicit, in itself existence. The move from in itself existence to in and for itself existence is

---

[3] Our term sublation thus coincides with the Hegelian term "Aufhebung," *Werke,* Vol. 4, *Die Logik 1,* p. 120.
[4] Our notions of existence "in itself," existence "for itself," and existence "in and for itself" draw inspiration from Hegel's account. See *Werke,* Vol. 8, *System der Philosophie 1: die Logik,* #96, Zusatz #223, etc.; and Vol. 10, *Philosophie des Geistes,* #387, 411, etc. An excellent discussion of these notions is given by Klaus Hartmann, *Sartre's Ontology* (Evanston, 1966).

a move to a categorial richness that contains its contrasting term as a moment of this richness.

The moment of the dialectic that expresses the higher category's relations to the lower category is termed the "regressive dialectic"; it is the movement to the necessary grounds presupposed for the possibility of the higher category. The other moment of the dialectic is the progressive dialectic – to quote C. S. Peirce, "that deep study of each conception in all its features [which] brings a clear perception that precisely a given next conception is called for." [5] In terms of an account of appearances, dialectical progression is the further development and fulfillment of one type or category of appearance in another but higher type or category.[6] It is the affinity manifested in a lower category's implying its fulfilling category. It is the sense of progression up the levels of significance presented in an ontology of being. It is the resolution of the incompleteness of lower categories.

2. Characteristics of the dialectic

Consideration of the dialectic takes us back to the issues which were touched upon in Chapter I C2, and to the nature of thought itself. Thought is that in which the significance of the eidetic structures of being is manifest. Other modes of consciousness when prescinded from thought do not encounter the categorial structure of being. A feeling, for example, divorced from thought does not recognize its content as a certain type, nor even as an object. In feeling as such, there is no object nor possible distinction of object and subject. Feeling encounters the categorial, but this is apprehended only in thought. To endow feeling with recognition of the categorial is to transmute feeling into thought; to deprive thought of the recognition of the categorial is to render it mere feeling. Only in consciousness unitable under an "I think" can an object be distinguished from a subject, for only then does the possibility of the recognition of structures and thus the possibility of experience arise.[7] Or in ontological terms, appearance is more than an object for feeling. It has a sense of independence and objectivity; it is an object for knowledge and thus for thought.

---

[5] *Collected Papers of C. S. Peirce*, 1.490.

[6] Hegel finds this description applicable to the higher stages of the dialectic. "The progress of the notion is no longer a going over or a reflection into something else, but a development." Vol. 8, *op. cit.*, #161.

[7] Such a distinction between feeling and consciousness is delineated by Hegel. See *Werke*, Vol. 10, *Philosophie des Geistes*, #403, 404, and 413, 417.

Experience, as the encounter of objects and thus of a world transcendent to any *particular* subject, is dependent upon the presented distinction of object and subject. The particular conscious subject must be able to find the object as one not dependent upon him alone for its existence. Otherwise, the potential intersubjectivity of objects is not realized and consequently neither is the objectivity of the world. This objectivity of the world is presented in the eidetic structures, for they are what must be confronted by any particular subject who confronts that type of world. It is in terms of these eidetic structures that there are objects of certain types, particularized in specific fashions, and presented in certain ways. They are the universality of objects; their existence is for every man. The significance of being is a significance for thought, and objectivity is constituted within and through its categories.

Finally, consciousness can bend back upon itself, become self-consciousness, and make itself its own object.[8] Through this turn it can reveal the eidetic structure in thought of the unity of the categories of appearance; it can render its own structure thematic. This makes possible the completion of an ontological account by passing to the articulation of the categories in the unity of thought. One can attempt to discern their relations to thought as a whole, as a totality. Thought as this totality will be invoked as a goal with reference to which categories can be ordered as stages in the achievement of this goal. (See Chapter I C2 and III B.)

Thought (as reason) is explanation that moves towards a self-explanation that would explain the content of thought to itself and leave no surd residual elements. Reason has no rational limit, for as a pure principle it is without qualification. It is posited as absolute.[9] This is an eidetic description of reason; any limit upon reason is not integral to it as such, but to the idiosyncratic stance of a particular subject. Moreover, reason is posited as an absolute ground of the categories, since being has significance only in relation to mind [10] – thought is the ultimate ground of the significance of being. It cannot be otherwise, for i) mind is that alone which intends; thus, only in experience is meaning presented; ii) all predicates of being are notes of significance in experience, and apart from their possible reference to a mind they are devoid of significance; thus,

---

[8] Hegel likewise describes self-consciousness as this further stage of consciousness. See *Werke*, Vol. 10, *op. cit.*, #417 and 424-427.

[9] As Hegel said, "The absolute idea alone is Being, imperishable Life, Self knowing Truth, and is indeed all Truth." *Werke*, Vol. 5, *Die Logik 2*, p. 328.

[10] It should be quite clear that mind in this sense is "reason." Or one could say, mind is reasonable. It possesses the moment of "in-and-for-itself-existing universality and objectivity of self-consciousness" which is the truth of self-consciousness, no longer considered individually, but universally." *Ibid.*, #437.

iii) any assertion of the being of anything implies a reference to a possible experience, and iv) being itself can only be referred to in relation to possible experience and thus unification with thought. Consequently, v) thought embraces all the possible categories of being. Thought exhibits the richness of meaning in an inference that is never of its own nature terminated until all categories of significance are encompassed.

Thus, the final ground of significance is mind, for in terms of mind all else is significant and mind itself is relative to nothing but itself, since there is no ground of significance outside of mind. Mind grounds itself because its significance (as possibly grounding all the categories) is, as are all the categories, significant only in relation to mind. Mind alone is self-explanatory.[11] In mind the regressus towards significance ends, or rather becomes circular.[12] All being is significant in terms of being significant for mind, and mind is significant for mind as the stance in relation to which all other categories are significant. This is the absolute status of mind. As Husserl described it, "Nature is the field of thoroughgoing relativities and can be such because these relativities are always relative to an absolute which consequently is the bearer of all relativities: mind." [13] It is the category of the full encompassment of being in thought.

In virtue of this self-grounding status we describe mind as existing *in and for itself*. The movement from *in-itself* existence to *in and for itself* existence is the dialectical development of full reality. To quote Hegel, "In order to grasp what development is, one must distinguish two kinds of – so to speak – states. The first is what is known as predisposition, capacity, existence in itself (as I term it), *potentia, dynamis*. The second determination is existence for itself, actuality (*actus, energeia*)." [14] This progression from potency to actuality can be recognized in the relation between that which is merely significant for something else and that which is significant and recognizes its own significance. Objects other than minds can then be said to exist *in themselves*, but not truly *for themselves;* they exist *for mind*, their significance is grounded in mind. Consequently, their significance is only implicit in them but (potentially) explicit in relation to mind. They are in this respect alienated since their

---

[11] Mind as such is "the idea that thinks itself, the knowing truth." Hegel, *Werke*, Vol. 10, *Philosophie des Geistes*, #574. It is in this sense that absolute mind expresses a self-identity; see Vol. 10. #554.
[12] "The Notion which conceives of itself." Hegel, Vol. 5, *Die Logik 2*, p. 553. We share with Hegel this notion of the self grounding nature of thought which is essential to his philosophy. See Vol. 4, *Die Logik 1*, p. 75; Vol. 8, *System der Philosophie 1: die Logik*, #17; Vol. 10, *Philosophie des Geistes*, #574, 575, 576, 577.
[13] *Ideen, Zweites Buch* (The Hague, 1952), #64, p. 297.
[14] *Werke*, Vol. 17, *Vorlesungen über die Geschichte der Philosophie*, p. 49.

own significance is not *for them*; their being is sundered between themselves and mind.[15] Only in mind is this tension overcome through mind's being the realization of this meaning *in and for itself*. Mind embraces its other in itself by explicitly having the significance of its other integral to its own concreteness. Mind has as its content the world it knows, especially the relations of its body to the world. And by having this significance exist for itself, the mind is the actuality of the significance which the body is only potentially (e.g., the body appears as if it were intelligent, the mind is intelligent).

In general terms, mind is the point or stance where being becomes fully explained, or embraced categorially. That is, mind (considered as such and not as any particular idiosyncratic mind, see Chapter I C2) is *the* ground of the significance of objects in general; it is, as we have indicated, the category of the full relation of being to thought. Only in this posited stance are in itself and for itself existence fully coincident.[16] And in comparison with this stance all other categorial plateaus constitute various but incomplete stages of being's realization of self-explanatory significance. As the category of the full encompassment of being in thought it is *the* self-explanatory category. As such, and insofar as reason seeks that which is translucent to thought, this category or stance must be the *terminus ad quem* of reason. Or more fully, in order to understand i) the relation of being to thought, and ii) the relation of reason to the categories of reality, we posit a categorial focal point which expresses i) being grounded in thought and ii) reason's nisus towards more encompassing explanation. It is a posited final direction of being and thought that serves not only to explain their relation but to allow us to articulate the categories in a particular order.

We can then articulate categories in terms of this final stance. It functions as a telos towards which each categorial level can be oriented so that the categories form a notional system and not an aggregate. In terms of this stance, reason, as thought's drive towards the full comprehension of being, requires that for any one incomplete category there must be a

---

[15] Their alienation lies in their being a mere something-for-another and thus is always involved in otherness "insofar as what something is in-itself belongs to it, it is infected with being-for-another." Hegel, *Werke*, Vol. 4, *Die Logik 1*, p. 141.

[16] Hegel remarks, "Absolute mind comprehends itself as that which posits being as its own other, bringing forth nature and finite mind. Thus, this other loses every appearance of independence from mind, completely ceases to be a limit for mind and appears merely as the means by which mind succeeds to absolute existence for itself, to an absolute unity of its existence in itself and its existence for itself, the unity of its notion and its actuality." *Werke*, Vol. 10, *Philosophie des Geistes*, #384, Zusatz, pp. 37-38.

next more complete category which embraces the lower category as the necessary moment of its higher truth. Any particular category can be judged as falling short of full encompassment of categorial reality, and any one particular category requires a next fuller, more encompassing category that would be the next stage in the move to complete comprehension of the richness of being. Richness here refers to the degree to which a category of reality is self-explanatory, the degree to which it embraces the principles required for its own explanation. For example, mind is a richer category of being than body; body requires mind for its explanation while mind embraces body as a moment of its existence. This ordering according to richness or encompassment is an expression of thought's commitment to explanation. The founding-founded structure of the categories is transfigured by this nisus towards complete explanation. A systematic articulation of the categories is attempted as a program for rationally appreciating the various categorial levels of reality; categories are posited as stages of explanation in order to understand them in terms of a central goal of reason. The categories are then united in a dialectical structure that constitutes a ladder of ascending significance, with thought as the inference of explanation that joins them one to another on the road to complete categorial encompassment.[17] This is desirable since it allows one to justify categorial relations, to say *why* they exist. One understands categories as fulfilling reason's demand for a further stage of categorial richness. Moreover, insofar as one recognizes the grounding of being in thought, this is unavoidable. To be encompassingly, not onesidedly, significant, being must realize a diversity of significance, and this diversity must *eo ipso* be understood in terms of its relation to completeness. The absolute stance of mind as the final goal of reason vindicates without qualification the ordering of the categories in terms of the completeness. It presents a final and essential relation of being and thought. But it is enough for this investigation if one grants the project of giving a linear relation of categories that i) captures the dynamic nature of thought (i.e., its progression towards more complete explanation) and ii) satisfies reason's urge to accommodate being to its own purposes. As such it is a modest attempt at categorial analysis and the notion of absolute mind acts merely as a point of orientation in the ordering of categories.

The absolute status of mind is thus a prime principle of categorial ex-

---

[17] "The links in this chain are the individual sciences, each of which has an *antecedent* and a *sequent* – or stated more exactly, each *has* only an antecedent and shows its sequent in its conclusion." Hegel, *Werke*, Vol. 5, *Die Logik 2*, p. 351.

planation. The term "absolute" is meant to emphasize the priority of thought and to indicate that this priority is to be understood apart from the thoughts of any finite mind as such. Mind as Absolute is the posited notional articulation of the elements of thought prescinded from infection by the idiosyncratic. It is a category of totality: the total categorial encompassment of being. It provides the terminus of a dialectical progression of meaning through which categories in their diversity are ordered towards a single goal, integral to them. This is but to say that the categories of reason are to be understood in terms of reason as a whole; and that reason as a whole is to be understood in terms of a final categorial articulation and compass of reality – its consummate goal. Again, this absolute stance should not be thought repugnant because of metaphysical implications. It is invoked merely to specify the goal of reason – to give the direction of the dialectic.

Further, thought as a posited ground does not create being. Though there may be a next higher, more explanatory category of being, it need not be actually present. For example, a cosmos devoid of mind is possible, and has probably existed, though all along it was incomplete because mind and its further strata of significance were absent. The affinity of the categories is first and foremost a notional one. It expresses not what is, but what should be. It is ontologically normative, not ontogenic. It defines how the categories should be related but does not generate new levels of being. In order to know what is actual, one must resort to eidetic phenomenology.

Finally, it may be a rational psychological fact that unless one were acquainted with the *de facto* founding-founded structures of appearance, one would not be able to "see" any particular synthetic leap from a lower to a higher category, required by transcendental thought. Thus, in our investigation we proceed to a dialectical grounding of the categories of mind and body only after we have given a phenomenological account (i.e., of the *de facto* founding-founded structure of mind-body) and have discovered its incompleteness. Strictly, though, this psychological limitation is not relevant to the transcendental necessity of the dialectic.

In summary, the dialectic is the transcendental relation of categories. It is their justification in terms of a posited ground of being, which is also reason's goal of full ontological explanation, the stance of full categorial comprehension of reality. Thus in addition to the relations of a) presupposition, and b) tendency, which are apparent on an ontological level (see Chapter II A7), it exhibits a notional affinity and development of categories which can be characterized in three guiding principles:

i) A category tends to a higher category that explains it in a fuller fashion and within a new stratum of significance. This is the dialectical move to greater categorial encompassment. Encompassing more, the higher category is more *self-explanatory* and thus less incomplete. The higher category contains the ontological structure that adds a new kind of significance to a previous ontological structure, thus explaining the lower category's "implicit" significance. The second is implicit in the first, though only in being able to complete it in a categorially new fashion. The movement is ampliative. In this fashion the first category is explained by the second; its further significance now becomes manifest. As a moment of the second category the first need no longer point beyond itself for explanation; as an ontological unit a degree of self-explanation has been reached.

This move from the implicit to the explicit must not be construed analytically or its ampliative nature will be lost. That which is implicit is the possibility of a further determination of being that would constitute a richer notion of being. The move from the lower category to the higher category must constitute a step in reconstructing the significance of the higher category. This move is, as we have already noted, guided by the phenomenological requirement of encompassment (see Chapters I C1c, III A4 and B, and immediately above); in other words, its claims are not arbitrary but guided by appearance. The reference to appearance is, though, transmuted insofar as it is theoretically acknowledged. One is not concerned with the founding-founded structure of appearance as such; one has already examined that. For this reason the question of correspondence with reality is already settled; or more properly, due to present transcendental concerns it does not arise. One instead notes the ordering of categories in terms of the nisus of reason towards full explanation. In this regard the relation of body and mind is paradigmatic. Mind is the content of reality in the context of self-explanation (i.e., as discussed above, self-grounded). One knows phenomenologically that mind is the richer category. Here, though, one focuses on the relation between two structures with reference to reason's need for explanation. One then recognizes the second category as going beyond the limits of the first, answering to richness which the first could not contain. The second is thus a more embracing ontological principle; it specifically answers needs raised by, but unfulfilled by the first category.

This move is progress towards constituting a richer category of being (insofar as one refers back to an initial phenomenological description of reality, this move could be termed a notional reconstruction of the presented structure of reality) and towards a categorial grounding of explanation (i.e., a move towards categorial self-explanation). In other words, one moves towards a category that explains the features of the first category which were beyond the explanatory scope of the first category, but to which the first category in defining itself made reference. It is in this sense that the second category realizes an explanatory scope only implicit in the first.

This is a notional development of the phenomenological relationship of "tendency" (vide Chapter II A7). Phenomenologically we could only indicate that *de facto* certain categories were naturally embraced by others. But now in terms of the demands of reason we are provided with a principle for positing a next category. A next category is posited as the higher truth or explanation of the previous category. The positing captures the dynamism of thought: the movement from a less adequate to a more adequate categorial stance, from a less to a more illuminating notion of reality. This is the dialectical principle of *full explanation*.

ii) The principle of full explanation can be viewed in reverse. Instead of attending to the relating of categories in terms of the telos of greater richness, one can view the *vis a tergo:* the movement away from the failure to encompass richness. Phenomenologically the symptom of such categorial deficiency is an ontological instability. A category when viewed in its completeness may begin to show elements of significance proper to another category. For example, the complete body acts "as if" it were ensouled; this, though, involves a reference to elements that fall outside of the category of body: This pointing beyond can be appreciated in terms of the needs of reason – thus transcendentally. The category is unstable because its unity becomes involved in elements of significance, which it cannot incorporate; it attempts to be encompassing, breaks down, and requires a further attempt at unity. The conceptual deficiency acts as a motivation for a further category just as the goal of full explanation does. The difference is one of emphasis. Here the disquietude over disunity and failure to encompass is central. A category is forced to relate to its opposite in order to account for itself – there is antithesis and diremption of a claim of encompassment (e.g., mental terms, which contrast with body terms, become apposite for the explanation of body, disrupting the provisional completeness of the category body). The tension between the category and the contrasting but requisite elements outside of it impels us to search for a new categorial unity. That is, when a particular category fails to encompass an explanatory principle that would particularly illuminate its content, thought must posit a category that embraces the previous category and the excluded principle. This is the dialectical principle of *ontological stability*. Along with the principle of full explanation it constitutes the dialectical completion of the ontological categorial relation of tendency. They are the principles of the progressive dialectic.

iii) Any category, as concrete, presupposes a lower category as its necessary moment, not merely because that category is necessary to its significance, but because the higher category is the further development of the lower one. A higher category must have a lower category which it sublates and possesses as a necessary moment. That is, since the dialectic is a synthetic implication, it presupposes its protasis without which the apodosis loses its special explanatory significance. The full character of a category is found not only in reality but also in its explanatory relation to other categories. This principle will be termed the *principle of concreteness* or the principle of the regressive dialectic. It is the dialectical completion of the ontological categorial relation of presupposition. The presupposed category becomes the explanandum that is necessary precisely because of its relation to its explanans. A *de facto* relation is justified in terms of its supplying reason with content. Reason recognizes that a higher category owes its distinction to the lower category that it completes.

This has an implication for the interdetermination of ontological structures. The lower category as presupposed by the higher is a structural *sine qua non* for the higher. Its structures predetermine the higher structure – it is that which the higher must complete. The higher structure for its part adds a further dimension of structure and significance, but all along requires the previous categorial dimension so that it can then add a further depth and unity of structure. As Hegel suggests in the Zusatz to Section 96 of the *Encyclopedia*, the category of mind has truth, functions as an explanatory principle of being, only through reference to the body. The ideality of mind, its role as a higher order categorial principle, presupposes the reality of body, the lower order principle, which it elucidates. The concreteness of the higher category is its fuller explanation of the lower category; the higher category has the lower category as its appropriate content.

These principles can be further understood through employing the ideas of unity, diversity and affinity. A category is a general conceptual unity ingredient in the structure of appearance and integral to the concept of being. It encompasses certain diverse elements of significance. Insofar as this unity succeeds it is the rationale of the coherence of these elements, it is the basis of their affinity. But insofar as the concept is not fully encompassing, insofar as general elements of the significance of appearance fall outside of its compass, it is onesided. Moreover, the elements outside of the unity define the category as well as its onesidedness. The definition of any particular category involves setting limits to the category (*definire*, to set limits). Consequently, the category becomes understood in terms of its other, of that which it is not: the significance defining and limiting the first category (e.g. body as not mental, non-intentional, etc.). This further significance must be understood conceptually and thus within a new unity. It is significance potentially recognizeable as elements in a further categorial unity opposing the first category (e.g. the mind as the non-physical); the further unity defines and thereby delimits the first category (e.g. the body as the non-psychical). But in defining the first category, the second must take account of the first (the body as the incarnation of mind) if it is not to be as onesided as the first. The first category must be rendered not just an other to the second, it must become a moment of the second (e.g. the body as the embodiment of mind). The other to the first category consequently emerges as a definite other, a determinate negation: the first category's unity developed to encompass that which previously defined it and its limits. The second negation (the determinate negation) negates the onesidedness (the first negation) operative in the definition of the first category. The second category is developed in order to satisfy reason's interest in a more encompassing, more self-explanatory category through resolving the first category's incompleteness within a new, more complete, more encompassing categorial structure. Part of the function of the affinity of the elements of the new category is to embrace the previous category as an element or moment of a new unity. This involves reference to the goal of complete explanation and thus contains a reference to the nisus of the dialectic, as well as to its *terminus a quo* (a less encompassing categorial apprehension of being) and to its *terminus ad quem* (a fully encompassing categorial apprehension of being).

The drive to overcome onesidedness and ontological instability (the need for a category, an ontological principle, to refer beyond itself in order to account for its own function as a principle), leads to the accumulation of ontological concreteness. To achieve a more self-explanatory, self-contained principle of being, a richer category is ordered after the less rich category – towards the specific end of including the first category within the second category through the second category presupposing and thus giving an account of the first. One has then a series of categorial unities distinguished in terms of encompassing more or less of being's significance. The elements of the unity cohere in an explanation, a new category, sought in order to find a further, more explanatory determination of the general category of being. The concreteness of ontological principles becomes a *result* of reason's needs and reason's operation. Because of this, the concreteness can be said to be *understood*. It is important to remember that this is the motivation for employing the dialectic as a means of ordering and relating categories. One does this to *comprehend* the relation between various categories – such as mind and body.[18]

---

[18] Hegel's account of the dialectic is much more detailed, and it is to this that our presentation is most indebted. The leitmotifs of our analysis of the dialectic are, though, drawn from Kant's systemic principles of unity, diversity, and affinity (see *The Critique of Pure Reason*, "The Regulative Employment of the Ideas of Pure Reason").

Hegel outlines the general structure of the dialectic as well as the mechanism of individual dialectical movements. His sketch of the dialectic indicates that the dialectic is:
1) a principle for *ordering categories* from the least to the most determined category. "This progress is primarily determined towards this end – that it begins with simple determinations and those that follow become ever more rich and concrete." (Hegel, *Werke*, Vol. 5, *Die Logik 2*, p. 349.) "Every new level of going outside itself, that is, of further determination, is a going inside itself; the further *extension* is as well a greater *intensity*." (Vol. 5, p. 349.)
2) A principle for *accounting for categorial structure* via a logical generation of categories. *Categories are rendered the result* of concrete reasoning (i.e., consideration of the matter of categories – determinate negation). "Such a negation is not all negation but negation of a determinate subject matter, which dissolves itself and thus is definite negation. Thus the result contains in essence that out of which it resulted. This is really a tautology; but if that were not the case it would be something immediate, not a result. In that the result, the negation, is a determinate negation, it has a content. The determinate negation is a new concept, and a higher richer concept than the one which preceded it." (Vol. 4, p. 51)
3) A principle for *explaining* concrete reality *through reconstructing* reality in conceptual terms. This involves construing being in terms of thought. The *Logic* "is a reconstruction of the thought determinations which are thrown into relief by reflection." (Vol. 4, p. 31) "Philosophy gives the content of the empirical sciences the essential form of the freedom of thought – that is, an a priori character. The content is authenticated by necessity, in place of being verified by finding and experiencing a fact." (Vol. 8, #12, p. 58-59.)
4) A principle for *justifying categories through* placing them in *a system* that satisfies reason's need for the self-explanatory. "A philosophy which is not a system cannot be scientific... A content has its justification only as a moment of a whole. Outside of a system it has only an unfounded presupposition or subjective certainty." (Vol. 8, *System der Philosophie 1*, #14, p. 60.) "The logic is itself the pure notion that has itself as its object. It is the notion which as its own object runs through the totality of its determinations and develops into the whole of its reality – into a system of science. It concludes with this comprehending grasping itself..." (Vol. 5, p. 352.) Our principles of self-explanation and concreteness include these principles, but more globally and less precisely since we are not immediately concerned with the more systemic goals and nature of the dialectic.

Hegel's treatment of the different types of individual dialectical progression is in a sense the substance of *The Science of Logic*. There are, according to Hegel, three genres of categorial unity: being, essence and notion. Each has its own variation of dialectical progression consequent upon the type of categorial unity that has been generated – the nature of the determinate negation is a function of the type of Fürsichsein, categorial unity or inclusiveness, it can generate.

Before passing to the examination of the dialectical relation of mind and body, it is worth noting that these three principles (i.e. of full explanation, ontological stability, and concreteness) are not independent. Rather, we have isolated three illuminating perspectives on the dialectic so that we can use their viewpoints as methodological principles. The

Being, essence, and notion differ as to the unity and affinity of their conceptual determinations. These are "the basic elements, the unity of the notion itself and consequently the inseparability of its determinations..." (Vol. 4, *Die Logik 1*, p. 61.) In being (i.e., within the sphere of being delineated in the first book of Hegel's *Science of Logic*), being receives distinct determinations that are related to an other by passing over into it. "In the sphere of being, the finite transmutes itself and becomes an other." (Vol. 5, p. 104.) The sphere of being is thus a field of contrasts between determinations; "in the becoming of being, being is the ground for determination; the determination is a relation to an other." (Vol. 4, p. 493.) In being, then, unity is minimal, diversity preponderant ("the notion is posited in its distinction," Vol. 4, p. 61) and affinity is achieved through an ought (*Sollen*) transcending onesidedness via a passing over into an other. The dialectical transitions tend therefore to be abrupt. That is, a category is delimited and consequently has a barrier to completeness in its very determination. This barrier is the negation implicit in the category's significance as delimited vis-a-vis an other. This conviction of onesidedness engenders an ought – a need to transcend the barrier for the sake of overcoming a finite, a restricted perspective. The goal is a true infinite – a concept that includes the rational principles required to account for the concept in its concreteness. It is "the fulfilled ought, reflected into itself; it is being that is completely affirmative, solely related to itself." (Vol. 4, p. 160.) But there is not sufficient conceptual machinery available on the level of being to allow the completion of such a unity. "The categories of being were, as notions, essentially these self-identities of determination within their barrier or their being-other." (Vol. 5, p. 38.) Difference, not affinity, is prominent. There is no inward reflection on internal conceptual substructure that could render determinations into relata within an embracing unity. Determinations are therefore related primarily through confrontations, hence unidimensionally.
When the conceptual unity not only grasps determination but also the relation between it and its other, the level of essence is reached. There is a given supporting unity or identity of elements. "There results a sphere of mediation, the notion as a system of determinations of reflection." (Vol. 4, p. 61.) A new dimension enters, that of reflection, which provides a new axis of unity; inwardness that allows a systematic grasp of the determinations of being. In essence the limit of a determination no longer separates two starkly contrasting determinations. "The reflective movement, in contrast, is the other as the negation in itself, a negation that only has being as a negation that is directed to itself... thus here the other is not being with a negation or limit, but rather the negation with a negation." (Vol. 4, p. 493.) That is, the negation, the act of determining being, which involves delimitation and thus otherness, is muted by a second, mediating negation that apprehends the determination not as a naked other, but as part of an essential structure. "The determination of reflection has, in contrast (to being), reabsorbed its being-other. It is being-posited, negation, which bends back into itself the relation to an other. It is negation which is self-identical, the unity of itself and its other." (Vol. 4, p. 504.) The function of limit has thus been altered. Being in itself and being for itself achieve an explicit unity; "essence is an absolute unity of being in and for itself." (Vol. 4, p. 483.) This "essential being-for-itself... is the self sublation of being-other and determination." (Vol. 4, p. 483.) That is, through "relation to its unity" (*Ibid.*, p. 484), the conceptual unity of essence, a distinction is posited between determinations of reflection – which is to say all determinations are internal to essence. What happens then is that instead of sequentially introducing new

same can be said of the distinction of the regressive and the progressive dialectic. Indeed, the progressive dialectic is the dialectic proper, since it represents the synthetic movement to a higher category of meaning. The regressive dialectic is merely an analysis of the notion of a higher category in its concreteness. The analysis begins with what the dialectic has produced and thus does not itself involve the dynamic progression of the dialectic from a lower to a higher stratum of significance. We begin with it because of its similarity to the phenomenological analysis of the founding-founded structure already described in Chapter II.

determinations, in essence new determination-relations are introduced. This marks a new type of unity – an internal one giving a substructure to the field of determinations.

Finally, in notion, even the contrasts between determinations of reflection are overcome. They are no longer merely posited; this externality is replaced by a comprehensive, substantial, *sui generis* conceptual unity. It can account for itself in its own terms, be self-explanatory; it is a totality of significance, "the substantial identity." (Vol. 5, p. 31.) Thus essence progresses beyond being by positing determinations and relating them, while in being they simply are; the advance of notion is the positing of determinations, to the positing of the unity of the relating. "In the sphere of being the finite transmutes itself and becomes an other. In the sphere of essence it is *appearance* and is posited so that its being consists in an other shining into it. The *necessity* is an *internal* relation that has not yet been posited as such. The notion, though, is this: the *positing* of this identity; that which is, possessing not an abstract but a *concrete* identity with itself, being immediately in itself the being of an other." (Vol. 5, p. 104.) That is, the unity of the relation itself becomes thematic and is explained. The unity is no longer given as in essence. "It is no longer blind identity, that is, an internal one, but essentially has the determination." (Vol. 5, p. 11) In this unity, distinctions are developed as explicitly homologous with thought. In being and essence a heterologous element was still present, an implicit contrast between the conceiving and the matter of the conception. It is this that Hegel draws our attention to through the distinction between the objective and subjective logic. The objective logic preserves an analogy to the contrast between a concept and the object of the concept in the externality of determinations in being and essence. In the subjective logic, the logic of notion, the distinctions continue into and are in identity with what is different from that (Vol. 8, #240) – "only that is posited which is already present." (*Ibid.*, #161 Zusatz, p. 355.) The limit is progressively transformed, becoming more and more identical with and understood through the categorial unity.

Being, essence, and notion thus represent a progressive integration i) of conceptual significance (determinations – hence distinctions) ii) within a conceptual unity (Being: determinations taken as such – without a sub-structure. Essence: determinations taken in terms of their relations – substructured. Notion: determinations as interidentified in terms of being determinations of thought – the substructure and what it substructures become a true system or coherence of determinations. See for example "Mechanism," "Chemism" and "Teleology" in *The Science of Logic*.) iii) by means of various genres of conceptual affinity (in Vol. 8, #240, the various types of dialectical relation are summarized. For being there is determination by, and passing over into, an other. In essence, relation to and showing into an opposite. In notion: continuity and identity with that which is different – a developmental relation.). Thus the dialectic has three genres: further determination as an other, reflection into an opposite, development into a less onesided form.

In fine, it is suggested that an understanding of the dialectic can be augmented through understanding Hegel's more detailed treatment of the dialectic as an exhibition of the conceptual integration of categorial richness – a passage from "being as the pure notion in itself" to "the pure notion as the true being." (Vol. 4, p. 60.)

A TRANSCENDENTAL ONTOLOGICAL ACCOUNT

B. THE DIALECTIC ON MIND AND BODY

1. *The regressive dialectic*

a. The necessary conditions for the concreteness of mind – or the body as the presupposed explanandum

In Chapter II we described the ontological structure of the category of finite mind, noting that to prescind from the significance of body was to obliterate the significance of such a mind. A mundane mind excludes the sense of being able to act upon the world in perception and action and is individuated as one mind over against others. We will not repeat that investigation but presume it as a basis for the present stage. Here we will examine the relation of mind to body as a relation required in the interest of explanation. This involves discerning the notional development involved in stages of categorial determination. The stages have already been exhibited phenomenologically, revealing a founding-founded structure: the body considered naturalistically is distinct from but integral to the significance of mind. As such, body has *the* significance that can be further determined and become mind. We start with a commitment to understanding categorial affinities. In Chapter II it was enough to demonstrate the presupposition of the body by the mind; it was a question of the structure of a presented significance. Here, we wish to assess the role of the category of body in the explanatory significance of the category of mind.

The mind is recognized to require an antecedent category as an explanandum only when the category of mind is understood to be a dialectical product. Only then can one see that it must possess an internal dialectical structure. This is circular, but not without warrant. We are, after all, examining mind-body not in order to discover its structure, but to discover the notional rationale implicit in this structure. The structure itself has already been explicated. Our goal at this juncture is to analyse out the dialectical antecedents which will allow us to reconstruct the eidetic structure of mind. That is, we start with mind as a particular categorial level of ontological explanation. The previous levels can then be treated as content for this explanation, since they are required as the basis for a movement to this further explanatory stance (i.e., mind).

The category of mind is, as such, a concrete, determinate structure of significance; it is a rich category.[19] It is abstract only when it is con-

---

[19] "The concrete is the universal that is determined, that is, which contains its other in itself. Initially mind is abstract. In this embarrassed condition it knows itself to be

sidered as opposed to nature and apart from its dialectical relation to nature – as, for example, in dualism or monism. Indeed, mind in its richest sense is thought grounding itself as the ground of all the categories (vide A above). This relation to concreteness also holds for any particular mind. Mind is that category of being which realizes what it is. Physical nature, on the other hand, does not know itself. Only mind exists, not just as an object, but also as an object for itself. But to be something for itself it must have a content to present to itself; otherwise it would be merely an idea. Minds are minds in being that in terms of which something is significant. Minds have a world in a way that objects and ideas do not, for in experiencing a world they, at least partially, realize and ground its meaning. They cause its potential significance to be actually significant. In this regard one can say, a bit poetically, that all minds are minds in the image and likeness of absolute mind. They have their concreteness, their meaning, through realizing the meaning of nature. But then not only is the significance of nature grounded in mind, mind presupposes this significance for its concreteness.[20]

Mind is the realization of what physical reality is implicitly, something to be known that does not know itself. In the case of absolute mind, this implicit status is posited as fully sublated and set aside categorially by the grounding of the significance of nature. Finite minds, on the other hand, cannot unequivocally be the ground of the physical world; they must be so only in bits and pieces. Since they are finite and thus not ultimate and complete grounds of significance, they must be finite in their realization of the world; they have a perspective upon the world that is one among a possible many. Otherwise, they would embrace the possible infinite number of spatial perspectives and temporal perspectives "at once" and from all "viewpoints" – which would be a form of actual infinity. Finite minds therefore embrace the world unevenly, both in respect to space and time. They appropriate the richness of the world in some places and in some times rather than others. As absolute mind is the posited idiosyncratic stance in which all significance is realized, so a particular mind is the realization of the significance of particular objects, a finite segment of physical nature. They are a certain portion of the physical world become mind (and thus "mine"!).

distinct from and in opposition to its other. But as concrete spirituality, mind... comprehends its opposite and does justice to it." Hegel, *Werke,* Vol. 17, *Geschichte der Philosophie,* p. 109.

[20] Hegel clearly recognized this presupposition of nature by mind. See *Werke,* Vol. 10, *System der Philosophie 3: Philosophie des Geistes,* especially #381, 384, 386. Also Vol. 8, *System der Philosophie 1: die Logik,* #96, 219, 223.

But mind is the special realization of only a particular category of physical objects. This category of objects must have the implicit significance that a mind uniquely makes explicit; it must be an instance of the particular sort of matter that can be best explained as an embodiment. The physiological parameters of the body satisfy this requirement. The body is, as described by physiology, a mechanism capable of discriminating general structures in its environment and responding to these structures in an equally (or more so) structured fashion conducive to its own homeostasis. Its "mental" performances are realized by no one but could be explained as intentions if only it were a mind. Thus there is an ontological tension between its behavior "as if" it were intelligent, and its physical reality. The body provides material that could be an element of a richer explanation. Indeed, mind does not just further determine the significance of the body; it is the translator (the *explanator*) of the body's reception and operation into perception and action. By being the category uniquely explained by the category of mind, body is the category uniquely required for mind's concrete reality. The body as this presupposed in itself existence of mind will be stipulatively termed the "physical analogue of mind." We wish to adapt the biological notion of analogue [21] to express the correspondence between a person's mental and physical reality, as well as to stress the mind's need for an embodiment of all its mundane reality. Mind must realize this category of body if, as the category of mind, it is to be a mind, not merely an abstract idea of mind.

This involves a dialectic of meaning in which the richness of the significance of mind is understood as the result of this further development of the category of body. The body is responsive and thus aware in itself yet not for itself; it is intelligent only for others. On the other hand, mind is that which is aware that it is aware; this constitutes an important further determination of mere in itself awareness. The relatively indeterminate significance of the body provides the mind with a foil against which and with which it determines its categorial concreteness. The body's mute meaning is that which mind gives voice to and indeed which provides mind with a perspective of reception and response to realize as perception and action. Consequently, the body is not merely *de facto* the founding structure of the mind's significance. It is also the category notionally presupposed for the concrete significance of mind. That is, if

[21] "An organ similar in function to an organ of another animal or plant, but different in structure and origin." *Webster's New Collegiate Dictionary* (Springfield, Mass., 1960), p. 32.

the category of mind is to have been posited as a next higher stage of categorial thought, there must have been a previous category that was everything but mind. It would have been precisely this note of being "everything but" that would have stimulated reason to posit a further category.[22] A previous deficient category is presupposed in the existence of a higher category. The mind is the discriminative and responsive complexity of the body existing in and for itself. Thus the body has an existence in itself that mind naturally brings to for-itself existence; as this for-itself existence, the mind presupposes the in-itself existence of body.

b. Summary

The regressive dialectic is an aspect of the notional explanation of the categorial relation of mind and body. Specifically, our analysis exhibits the body as the necessary condition of mind's concrete significance through

i. attending to the indeterminate category that must be presupposed if the more determinate category of mind is to be encountered, and

ii. thus understanding that mind must presuppose body as a necessary moment, since the body is the content which mind explains. It is that which is in itself so that mind can be in and for itself.

iii. This relation is one of meanings, though of meanings found ingredient in reality. It is a relation that establishes the structure of reality's significance; it explains why mind is encountered in a founding-founded structure of meaning in which body is the ground of mind's presence in the world. But it does not explain how this occurs – this is left to empirical psychophysiology and physiological psychology. Here instead we have asked why the category of mind should embrace the category of a body. Our answer was that the category of mind is not abstract and isolated from other categories. Rather, mind further determines the content of the category of body. It requires the body because it exists as the further explanation of the body's significance. The higher stage of meaning requires the lower stage precisely so that it can be a "higher truth." Otherwise the category of mind is onesided.

The dialectic of mind and body is thus not a special form of causality, it is not a physical dialectic.[23] Or at least if it expresses an empirical or metaphysical truth this question can be prescinded from the relation of meanings. We are concerned with this relation only as a dialectic of ideas

---

[22] The "everything but" is an index of ontological instability pointing out that one category of being requires still another for the full explanation of its content.

[23] See, for example, Engels' *Dialektik der Natur, Marx-Engels Werke,* vol. 20 (Berlin, 1968).

– a relation between the significance of categories.[24] This suffices, since we know already via our phenomenology that this structure of significance is a structure of the real.

2. *The progressive dialectic*

a. *The "as-if"-ness of bodies – incompleteness and instability*

We will now examine the notional progression from the category of body to the category of mind. Our point of departure is the body as a physical object, an identity pole of non-mental occurrences. As such it is an identity pole of electrochemical discharges and transmissions of energy in specific patterns in part corresponding to the patterns of energy stimulating the sensory receptors of the nervous system, an identity pole of initiations of specific limb displacements in response to specific states of affairs, etc. Other types of physical objects with other types of physical occurrences do not found the category of mental events. As an eidetic phenomenological observation, men, not stones, have pains. It would *appear that* the body's tendency towards being mind is integral to the type of physical occurrences of which the body is the identity pole. We will now examine these physical occurrences in order to explicate their relation to the category of mind.

A living body is, as a physical object, quite unlike other physical objects. i) It internally reproduces the patterns of its environment. Its sense organs function as transducers and translators, transducing photic, acoustic, kinetic energy, etc., into electrochemical energy and translating the patterns of received energy into patterns of electrochemical energy (and perhaps further translating these into patterned sequences of bases in RNA structures stored in the brain).

ii) Moreover, these introflected patterns direct its operations. Guided by them the body functions to preserve itself, to reproduce itself, to secure its own peculiar homeostasis, etc. Thus, it not only acquires an internal reflection of the world, it also behaves in a manner that imposes a certain form upon the world. Patterns of electrochemical discharges are transformed into mechanical operations that transform the patterns of the world outside the body. The pattern of this transformation is in turn itself received and directs further operations. Thus, these bodies tend to shape their environment, procuring energy sources and materials for

---

[24] For a short distinction of real and ideal dialectic see Hans Rademaker, *Hegels Objective Logik* (Bonn, 1969), p. 19-22.

self-maintenance, dominating and restraining destructive or harmful forces, and insuring the possibility of complete reproductive cycles.

iii) But assimilation, domination, and reproduction are not merely effected; their patterns are stylized and expressed in artificial patterns made to exist independently in the world. In particular, material is formed into shapes serviceable for certain types of behavior. Once they possess these forms their status is relevant not only for the body that produced them, but can be appropriated by similar bodies. In short, these bodies fit themselves with contrivances specially constructed for certain operations. These bodies transform natural objects into "tools."

iv) Certain internalized external patterns of nature (and patterns of operation indigenous to the body) are re-externalized in forms that influence whole groups of bodies "as if" they were external stimuli of a wholly different order. It is as if these artificial, external patterns stood for naturally occurring patterns of stimuli. Consequently, when one body stimulates another body with such artificial patterns of sound, visual shapes, etc., the other body responds to them as if they symbolized or signified other states of affairs. In short, these bodies construct "languages."

v) These bodies, moreover, compose stylized forms of interrelations and interaction in which different bodies assume different functions or roles. For example, certain bodies may gather materials that an entire group of bodies ingests. In addition, these stylized patterns of behavior a) do not arise directly from the external world, but through bodies introflecting its patterns, and b) the stylized patterns can outlast the existence of any or all of the bodies that first initiate these patterns of behavior. Other bodies take their places. These bodies, as it were, fashion "societies" with "institutions" and "customs." In short, the body, considered merely as it is presented in the naturalistic attitude (i.e., without reference to any mental significance), has the sense of being a complex universalizing mechanism. This significance distinguishes it as a unique category of physical objects, for unlike other physical objects the body internalizes and responds to the general structures of appearance.

We are forced to describe such objects as behaving "as if" they were intelligent, or in an "intelligent" manner. By this we do not mean to introduce the category of mind either precipitously or surreptitiously. This is rather an exhibition of the tendency or bias of such objects towards taking on the significance mind. The "as if"-ness or tendency represents an incompleteness in that the body is not completely understandable in its own right. In behaving "as if" it were intelligent, it reaches beyond itself. For example, if one secured an omniscient viewpoint and described a world in which there were complete automata, bodies that acted "as if" they were intelligent though they lacked intelligence, one would have to judge that they were singularly truncated.[25] Of course, one can only know of this appropriateness from the standpoint of a higher level. But from this higher standpoint (that of a mind committed to the goal of explanation) one can judge the inadequacy of the lower level. The "as if" behavior points to a stratum of sense which that world would lack and which nonetheless would be uniquely appropriate. This appropriateness of a higher and different stratum of significance indicates a synthetic leap from one level of significance to a higher level. That this leap takes place has already been established in our ontological analysis of the founding-founded of mind-body (see Chapter II). Our concern is now to determine *why* this should be the case.

Again, care must be taken not to interpret the "as if" mental status as a *petitio principii*. Our description of "as if" behavior does not covertly introduce the category of mind, but rather calls attention to the unique physical complexity of the body's structure and function. One must attend to this unique complexity to recognize the dialectical necessity that singles out certain objects of the world for the further and richer significance – embodied mind.[26] As merely a universalizing machine the body is incomplete. i) Though the body can "universalize" itself, discriminate itself as an object of a particular type, it cannot realize that it does so. ii) Thus the operations of discrimination occur but without the discrimination being anything *for* the body. The processes are merely *in* the body; the body exists as such merely *in itself*, not *for itself*. Thus, the body is, in our considerations here, merely an identity pole of physical processes, not an identity pole of intendings. Yet these physical processes discriminate general structures of the world in such a way that these discriminations can influence the whole of the processes identified with that one body. They are introreflected and guide the general operations of the body as a unit. But this status of the body is nothing *for* the body.

There is consequently a tension within the category of body. On the

---

[25] We assume for the discussion that such a state of affairs is possible. This is merely illustrative, though, for all that is required is that one distinguish the two strata of significance and apprehend the incompleteness of the physical stratum.

[26] The progression from the category of body to that of mind is unaffected by the possibility of describing the body teleologically. Non-intentional teleology adds to the instability of the notion of body: it is physical yet acts "as if" it were conscious of a goal. This emphasizes the contradiction between physical being and its "as if" significance that can only be resolved in the category of mind (i.e., the body, and its purposes, aware of themselves).

one hand it is recognizable as a universalizing machine. Neurophysiology and psychology interested in behavior patterns treat the body as competent to perform all the discriminative operations as physical occurrences that are imputed to mind as psychical events. Indeed, the nervous system appears as a united relatively autonomous physical system capable of receptive, perceptive, and expressive, etc., functions, so that "every action which grows out of it depends upon the working together of all portions of the brain. A destruction of any part of this total mechanism will cause defects in function of the whole cortex, and the severity of the defects will be proportional to the extent of the destruction. However, because of this functional and anatomic division, the symptoms will be more a reflection of defects in *understanding*, when the lesions are in the caudal half of the hemisphere; they will reflect defects in *doing*, when the lesions lie in the rostral half of the cortex." [27] Moreover, this mechanically self-sufficient stratum of physical significance is an eidetic prerequisite of the appearance mind (see Chapter II), as well as being integral to the concrete significance of that category (see Chapter IV B1).

On the other hand, this universalizing function of the body exists only *in itself* and *for us*, and not *for itself*, and is in this respect incomplete. The same content can be transformed so as to exist *in and for itself*. This tendency is the ever-present reality of a dialectical necessity: there is a next and fuller category of significance. This tendency in the category of body constitutes the ontological instability which is uniquely set aside by the category of mind. The impetus for this transformation is dialectical – it is the expression of the relationship between lesser and more explanatory categories (see Chapter IV A2). Without this dialectical move the body fails to attain its complete significance (i.e., as existing in and for itself); the body remains estranged from this fuller significance with merely a tendency towards it. The ontological instability can be recognized as a contradiction within the category of body. The contradiction is the tension between what the body is in itself but not for itself, and the tendency the body presents for us of acting "as if" it were aware and existing in and for itself. This would rephrase our description of the dialectic of mind-body in terms of the resolution of contradictions. The category of mind would be the overcoming of the contradiction in the category of body.

The possibility of this further category is guaranteed in general by the appeal to absolute mind; the significance of things exists only in terms

[27] E. G. House and B. Pansky, *A Functional Approach to Neuroanatomy* (New York, 1960), p. 430.

of mind, the ground of complete significance. Moreover, the category of particular mind is the realization of a particular body in a way analogous to absolute mind's grounding function (see Chapter IV B1). In short, the category of particular mind, the significance of a body existing in and for itself, is the natural further categorial move from body existing merely in itself, because in also existing *for itself,* the body becomes self-explanatory. The contrasting significance of the body is taken within the significance of mind (i.e., sublated) and mind makes it explicitly part of its own significance. The move is the dialectical move to further explanation and away from ontological instability.

In summary, the mind fulfills the sense of the body by transforming the "as if" sense of its behavior into an *is.* The stratum of psychical significance presents the body to us *as* self-aware significance; it is incarnate mind. In order to realize the complete or higher truth of its merely physical as-if behavior, body founds a higher stratum of significance, the category of mind.[28] The body is recognized as a moment in the reality of mind and what was considered as merely a body is given in appearance as incarnate mind. The unity of mind and body thus arises out of a body's fulfillment of its significance.[29]

b. The structure of the progressive dialectic

Having a considerable view of the progressive dialectic at hand, we can attend to certain cardinal features. Initially, we outlined two principles: full explanation and ontological stability. In practice they were intertwined. We noted that the category of body is incomplete; it indicates its further completion via a tendency towards, or betokening of, mind. This observation involves both principles. The category of mind is on the one hand the full explanation in which the significance of the body is more richly understood. On the other hand, it overcomes the ontological instability of the category of body considered in isolation: the body becomes a moment of mind. In the first case, we attend to the further enrichment of the category of body; in the second case we attend to the effect of considering the category of body in a more encompassing context. In both

---

[28] Hegel develops this notion of the mind as the higher truth of the body in his *Philosophie des Geistes* (*Werke,* Vol. 10), especially #381, 388, 389. Also vide Vol. 8, *System der Philosophie 1: die Logik,* #96, Zusatz, and Vol. 9, *Die Naturphilosophie,* #247, Zusatz.

[29] This assertion is meant to be an ontological, not a psychological, claim. The progressive dialectic is a categorial relation founded upon mind being the ground of thought. As such the relation does not rest upon either a psychological need nor is it primarily a question concerning the order of our knowledge of categories. It rests rather on the relation of different categorial levels in virtue of their differential encompassment of the richness of being.

cases we have attempted to indicate a notional relation between a less determinate, less complete, and a more determinate, more complete category. This relation can be adumbrated further to provide a concept of dialectical implication.

The biased category appears as a protasis reaching beyond itself. It is partially unexplained, since it could be the material for a categorially distinct level of significance; it is indeterminate but could be determined in a qualitatively novel way. The *implicans* thus does not analytically contain the *implicandum*. The category of body does not contain the significance, mind, but provides an introduction to mind. This leading into a new domain of meaning is terminated when a successful synthetic development occurs. Thus, the apodosis, the *implicandum*, announces itself when one finds the category that can determine the indeterminate significance of the *implicans*. The higher category is "implied" as the category fulfilling the lower category by giving the determination of meaning that renders it more self-explanatory. In the case of body, the mind appears as the higher category since it determines the in-itself existence of the body to exist in and for itself. It is the particular categorial completeness that a particular incompleteness suggests.

To some extent the discovery of a dialectical relation is a purely creative act of mind that defies characterization. It is pure spontaneity. But this does not mean that it is either surd or criterialess. Phenomenology has provided a sketch of the founding-founded structure for which we need only discover a notional warrant; the categorial structure is already given. The inspection of the lower category and one's encounter with its incompleteness is the beginning of the search for the dialectic. To recognize that the category of body has a bias towards being considered as mind indicates the direction of categorial development. This creative recognition itself can be understood as a nisus towards the stance of full explanation, the dialectical conclusion in terms of which all antecedent stages have their place. Thus to reveal a notional relation between categories, one must perhaps know the answer before one begins.

The dialectic provides an insight into the dynamic affinities of categories, which a strict phenomenological description leaves aside. By recognizing that mind is the dialectical conclusion of the body's significance, one at the same time comprehends why the category of mind should arise. To discover that mind is dialectically implied by body is to see that mind must be affirmed by the incomplete significance of body. It is affirmed as the new domain of significance that must be posited in order to fulfill the significance of body. Moreover, in exhibiting and satisfying the nisus towards more complete explanation, the dialectic relates the structure of mind and body to a final goal of reason: the full comprehension of being. That is, the category of mind *follows* that of the body since it is a further step towards complete categorial encompassment, a goal which the category of mind does not fully realize, but which it partly realizes and thus partly satisfies reason. Further, this intermediary stage is justified as a notional reconstruction of the concrete connexus of reality with which we began: mind-body. We initially confronted this reality and described its elements; now we have rearticulated it according to the notional affinities of these elements and their relation to the goal of explanation. The fabric of appearance is thus possessed as a dialectical result; its structure is justified. That is, in articulating the body and mind in an inference from less to more significance, one has demonstrated the rationality of the concrete structure of mind-body. The dialectical relation of mind and body is, after all, the necessary notional relation of the eidetic elements of appearance. It is the reason, the rational principle ingredient in this appearance.

c. An excursus: dialectical versus metaphysical objective idealism

Finally, we must emphasize the categorial nature of the dialectic. The dialectic concerns the logical development of domains of meaning, not the mere relation of different domains of appearance via certain generalizations or descriptions of *de facto* interdependencies. An examination of Peirce's objective idealism will illustrate by contrast the categorial concern of our investigation.

Objective idealism can be defined as the doctrine that i) there is no obstinate other to mind (this is the idealist aspect); ii) and that this fact does not make reality any the less objective. In particular, for C. S. Peirce objective idealism affirms two important propositions: i) matter is only a less explicit form of mind; ii) physical laws are based on cosmic habits. "The one intelligible theory of the universe is that of objective idealism, that matter is effete mind, inveterate habits becoming physical laws." [30] Peirce's development of an objective idealism is not merely the assertion of certain metaphysical postulates but the result of employing a notionally oriented methodology [31] for the construction of a cosmology. The application of such a methodology entails the rejection of ultimate surds, including explanations that are really only descriptions of ultimate parts.

---

[30] *Collected Papers of Charles S. Peirce*, 6.25.
[31] "Synechism is not an ultimate and absolute metaphysical doctrine; it is a regulative principle of logic, prescribing what sort of hypothesis is fit to be entertained and examined." *Ibid.*, 6.173.

Moreover, Peirce makes generality the final goal of explanation, because only generality is ultimately explicable, that is, fully conceptual.

Peirce's objective idealism (his assertion of the continuity of matter with mind) is founded principally on four considerations. First, alternative theories each possess a flaw. Dualism need not be considered unless simpler explanations fail (*entia non sunt multiplicanda praeter necessitatem*); neutralism which posits a monism based on an intermediate element likewise requires an unnecessary multiplication of entities. Materialism, on the other hand, fails because it cannot account for the occurrence of feeling; feeling and consciousness in a materialism remain inexplicable facts because mechanism explains nothing but the deployment of parts. Thus, objective idealism remains the best candidate if it can account for mechanical laws (this will be explored below). But this does not establish Peirce's objective idealism on its own merits, rather negatively through the inadequacies of its alternatives.

Second, because there is nothing in mere matter and pure physical causation outside of masses and their position in different instants of time, laws or universals transcend the bounds of a pure mechanism. As such, matter does not possess generality. To explain the reality of law in nature, Peirce requires nature to be essentially a structure of psychical events; they appear as physical (as will be discussed below) because they are viewed externally. Therefore, Peirce states that "analogy suggests that the laws of nature are ideas or resolutions in the mind of some vast consciousness, who, whether supreme or subordinate, is a Deity relatively to us." [32] This is quite different from saying that a categorial new stratum of significance must be recognized in which laws can be explained. This latter is a purely notional move.[33] The former is a metaphysical move concerned with the *physis* of things.

Third, a pure mechanism usually supposes a necessary development of the world due to certain arbitrary principles that obtained from the beginning of the world. Such explanations fail to be complete because of the arbitrariness of the necessary principles. Their inexplicability conflicts with the rational nature of laws. "To suppose universal laws of

---

[32] *Collected Papers of Charles S. Peirce*, 5.107.

[33] It is illustrated by Hegel's account of the development of the domains of chemistry and teleology from the domain of mechanics. See Hegel, *Werke*, Vol. 5, *Die Logik 2*, pp. 180-235, and Vol. 8, *System der Philosophie 1: die Logik*, #195-212. The relation is a categorial one. "But the necessary preliminary question is which of the two [i.e., mechanism and teleology] is the true notion, since they are in opposition? And the higher and proper question is whether there is not a third notion which is the truth of both of them or whether one is the truth of the other – the relation to goals shows itself to be the truth of mechanism." Vol. 5, *Logik 2*, p. 210.

nature capable of being apprehended by the mind and yet having no reason for their special forms, but standing inexplicable and irrational, is hardly a justifiable position." [34] Therefore Peirce proffers his theory of panpsychism and the evolutionary development of laws as habits of nature. Laws are regularities that have developed in the feelings of the cosmos. "... mechanical laws are nothing but acquired habits..." [35] It is upon these considerations that C. S. Peirce develops his cosmology in which the evolution of the universe is understood as the development of cosmic feelings into habits, of spontaneous acts of freedom into laws. This development is not a materialistic evolution, but an evolution of psychical elements, "... that in the beginning – infinitely remote – there was a chaos of unpersonalized feeling ... This feeling, sporting here and there in pure arbitrariness, would have started the germ of a generalizing tendency. ... the tendency to habit would be started; and from this, with the other principles of evolution, all the regularities of the universe would be evolved. At any time, however, an element of pure chance survives..." [36] Explicability is assured because the explication is made in terms of mental events which, according to Peirce, do not require further explanation. Thus, instead of a system of categories, Peirce proffers a cosmogony.[37]

Fourth, Peirce's account offers to explain how psychical phenomena are associated with physical states, without arbitrarily introducing psychical concepts. Peirce reasons that matter and purely mechanical explanations cannot suffice to explain mental phenomena, because mental phenomena possess more than physical properties. "Protoplasm certainly does feel; ... Yet the attempt to deduce it from the three laws of mechanics, applied to never so ingenious a mechanical contrivance, would obviously be futile. It can never be explained, unless we admit that physical events are but degraded or undeveloped forms of psychical events." [38] The physical laws and events of nature are thus, according to Peirce, habits and external manifestations of mind. Matter is rendered rational by objective idealism through considering it to be "effete mind." The distinction between mind and matter, though relieved of its troublesome ultimacy, is still preserved. Mental and physical being is distinguishable because of one's viewpoint. "Viewing a thing from the outside, consider-

---

[34] *Collected Papers of Charles S. Peirce,* 6.12.
[35] *Ibid.,* 6.268.
[36] *Ibid.,* 6.33.
[37] See Hegel, *Werke,* Vol. 10, *Philosophie des Geistes,* #577, in which nature is explained in terms of the categorial development of the Idea.
[38] *Collected Papers of Charles S. Peirce,* 6.264.

ing its relations of action and reaction with other things, it appears as matter. Viewing it from the inside, looking at its immediate character as feeling, it appears as consciousness." [39] The interaction of mind and matter is not that of ultimate opposites, but is accomplished through a continuum of mind and matter. "But all mind is directly or indirectly connected with all matter ... so that all mind more or less partakes of the nature of matter." [40] C. S. Peirce thus accepts a general correlation of mind and body in order to understand the essence of being. The correlation itself, though, does not develop out of the significance of these categories but is accepted in order to satisfy certain other theoretical demands.

Consequently, Peirce's account is metaphysical, not dialectical. He attempts to indicate the deeper *physis* of things that must be postulated to account for the regularities of appearance. Without regard to its truth, the account is incomplete in failing to explain the notional basis for the relation of the categories (see Chapter III B). As it stands, his account of the relation of mind and matter would be subject to a categorial analysis at the level of essence.[41] Such an analysis would leave the question of the phylogeny of laws and the appearance of matter, and turn instead to the dialectical relation of mind and matter as the inner and outer faces of being. Inner and outer, considered categorially, are correlative distinctions that are inextricably bound to each other due to their form; but at least initially they are indifferent to each other's particular meaning. They are related simply because they are inner and outer. "In Essence the determinations are only correlatives." [42] The significance of one does not develop into the other as is the case with the progressive dialectic of mind and body. Rather, in an attempt to comprehend the significance of the immediately presented reality of being, one posits an inner level. It is this structure that Peirce employs in order to understand the relation of mind and body. That is, their categorial structure provides the notional fabric of his account.

But to show that mind and body are the inner and outer faces of being is to offer material for a categorial analysis, not to give a categorial account. Peirce's account thus remains metaphysical; it claims first and

---

[39] *Ibid.*, 6.268.
[40] *Ibid.*, 6.268.
[41] "Essence is the simple permeation of the quantitative or external determination and of the inherent inner determining." Hegel, *Werke*, Vol. 3, *Philosophische Propädeutik*, p. 175. See Hegel's analysis of essence in Vol. 4, *Die Logik 1*, Zweites Buch, "Das Wesen," especially pp. 655-661; Vol. 8, *System der Philosophie 1: die Logik*, Zweite Abtheilung, "Die Lehre vom Wesen," especially #137-141.
[42] Hegel, Vol. 8, *System der Philosophie 1: die Logik*, #112.

A TRANSCENDENTAL ONTOLOGICAL ACCOUNT 119

foremost to explain the relation of reals through cosmological processes and structures. It only secondarily concerns itself with the relation of categories. Though i) one might wish to accede to Peirce's argument or even ii) assert that its conclusions are compatible with ours, still the basis for the assertions is distinct. Ours is based on a phenomenology and a subsequent analysis of the notional relations of the categories which this phenomenology reveals. His metaphysics would seem to be committed to the more general project of framing an a priori account of the ontogeny of reality. The first (a transcendental ontology) attempts to explain reality through explaining the notional relations among the categories of reality. The second (traditional metaphysics) attempts to explain reality through giving a theory of the real relations (e.g., material, formal, efficient, and final causal) among classes of existents. We have opted for the first, without the necessary exclusion of the second.

C. NEGATIVE AND POSITIVE DIALECTICS AND THE IDENTITY IN DIFFERENCE

Following Hegel's suggestion, we can distinguish both a negative and a positive dialectic.[43] This provides two artificially isolated dimensions of the dialectic, indeed of the progressive dialectic, which allows one to focus separately upon the dialectic as the denial of an incomplete perspective and the dialectic as the affirmation of a complete perspective of reality. The dialectic as negative is the dialectic as the apprehension of and denial of the incompleteness or onesidedness of a segment of being that has been isolated for consideration.[44] In the case of the mind-body relation, the negative dialectic is most particularly the denial of a physical monism as onesided and incomplete. More generally, it is the denial of the incomplete, vacillating dialectic between dualism and monism that has characterized the history of the mind-body problem. The vacillating dialectic is the unsuccessful attempt to acquire an ontologically stable account of reality: dualism collapses into monism, and monism collapses into dualism as each fails to give an encompassing account. Each instead simply contradicts its other while requiring it as a supplement for its incompleteness. The negative dialectic as the cancellation of this reverberating contradiction is the move from onesided metaphysical accounts to the encompassing transcendental ontology. It is thus also a negation

---

[43] *Werke*, Vol. 8, *op. cit.*, #81-82.
[44] This dialectical moment is, to quote Hegel, "the self sublating of such finite determinations and their going over into their opposite." *Ibid.*, #81, p. 189.

of the mind-body problem which is rooted in onesided analyses of mind-body (vide Chapter III).

The positive dialectic is the negative dialectic with a change of sign. What is denied by the negative dialectic is not the *reality* exhibited in a onesided view, but the onesidedness. The denial of dualism is not the denial of a real distinction between mind and body, but that this distinction excludes the possibility of a real identity of mind and body. Conversely, the denial of monism is not the denial of the real identity of mind and body, but only the denial that this identity is so ultimate as to exclude the real distinction of mind and body. So also for the relation of mental and physical monism. To deny physical monism is not to deny that the body is the ground of the mind's possibility; and to deny mental monism is not to deny that the mind is the higher truth of the body. Rather, the positive dialectic of mind and body affirms the relation of two distinct categories necessarily unified in a founding-founded structure based on the affinity of their significances.

This affirmation of reality by the dialectic has special significance for the relation of the categories of mind and body. The mind literally *realizes* the truth of the body. The mind is conscious of itself as an organic whole, not as a merely extended system of parts that behaves "as if" it were intelligent. The behavior is experienced and known by the mind. It recognizes itself as externalized, as physically present and active in nature. The mind is aware of itself as a unified corporeal whole living in nature and capable of persisting as this whole, as long as it is not so physically altered as to preclude sentient behavior. In this astonishing dialectical relation, the externality of physical objects and physical parts is overcome. The idea of intelligent behavior that was presented externalized in "as if" behavior is aware of itself; the mind, an intelligence, is aware of itself as being intelligent. The idea of intelligence has come into possession of itself.

But in all this, the material significance of the body is preserved – the dialectic as positive maintains the opposition of mind and body in the structure of a negation of a negation. It is the negation of the incompleteness of a lower category, which as incomplete is the negation of the higher category. But in negating the incompleteness of the lower category, the higher category preserves the lower as a necessary moment; the lower category is the original negation upon which the dialectical result depends for its positive character.[45] That is, the in and for itself existence

---

[45] "The dialectic has a positive result because ... it is the negation of certain determinations, which are just for that reason included in the result." *Ibid.*, #82.

of mind is a negation of the self-sufficiency of the mere in-itself existence of the body through encompassing in-itself existence as a moment of its own significance.[46] The negation of the negation is a determinate negation by possessing the first negation as the indeterminate content which it determines; the mind transcends the body by being conscious matter, not by being merely immaterial.

Mind and body thus remain each other's negation. They are others to each other and in this contrast they determine each other; each term of the contrast has its meaning in and through its other. The mind has its significance by embracing the significance of the body, and the body by having its significance fulfilled in the category of mind. This mutual mediation, though, has a direction which is expressed in the movement from the immediacy of in-itself existence to the richer determinacy and completeness of in and for itself existence. The concreteness, the richness of this completeness is the positive character of the dialectic which denies the abstraction and onesidedness of either the mind or body in isolation from the other. It affirms the affinity of significance that forms the basis of the organic connexus of the appearance, mind-body. The body must identify itself with the contrasting but completing category of mind, and mind must identify itself with its other, the body, the necessary moment of the mind's concreteness. This dialectical interidentification is an "identity in difference," because the contrast of body (existence in itself) and mind (existence in and for itself) must be preserved if the mind is to *be* the overcoming of the externality of the body, and not a onesided notion.[47]

This identity in difference of mind and body is the categorial structure which notionally validates the significance "my body." The notion "my body" requires both a coidentification and distance between mind and body within which the body can be singled out as me in the world and yet as an other to me. My body is mine because it (and not an other body) is the concrete moment of my presence in the world, while my mental life is the fulfillment of the same particular body. Moreover, while the

---

[46] "When I say that I am for myself, then I not merely am, but I negate in myself all that is other and separate it from myself insofar as it appears as external. It is negation by otherness, which negation is against me. Thus, being for itself is the negation of the negation. And this I term absolute negativity. I am for myself and negate otherness, the negative. And this negation of the negation is therefore affirmation. This relation to myself in existence-for-itself is thus affirmative. It is being that is its own result and is mediated through an other – but through the negation of an other. The mediation is preserved in this negation but as a mediation that is also sublated." Hegel, *Werke*, Vol. 17, *Geschichte der Philosophie* 1, pp. 383-384.

[47] Or as Hegel said: "Positive reason grasps the unity of determinations in their opposition." *Werke*, Vol. 8, *System der Philosophie 1: die Logik*, #82, p. 195.

embodied mind embraces this moment of physical existence, it also implicitly embraces the distinction in significance between mind and body. The identity pole of mental life and the identity pole of the body's corporeal reality contrast but coincide and are united in the life of mind, because the body is the most intimate content of the life of mind, though all along distinct from mind. The concrete existence of a mundane person is characterized categorially as an identity in difference.

An identity in difference is, as we pointed out in Chapter III, the goal of a complete account. In this chapter we have added the dialectical basis of this relation. Identity in difference is the form of the dialectic of being, since the dialectic is the rational unity of categorially distinct strata of reality; or as Hegel said, "Unity must be grasped as present and posited in diversity." [48] "The nature of speculative thinking ... consists entirely in understanding opposing moments in their unity." [49] By showing this dialectical basis, we have confirmed our phenomenological conclusion in Chapter II: the unity of mind and body arises out of their respective significances, out of mind's presupposition of body and the body's tendency towards mind.

The result of these considerations is a way of understanding mind and body, which is freed of the thing-conceptualization of most previous accounts. Mind is not a second thing posited over against the body, with its own independent matrix of causality and laws, which must then be brought into a unifying relation, monistic or dualistic, with the body. Mind is the body in its full reality – precisely no longer as mere body but as body cognizant and volitional: body as self-aware, body as-become-spirit. But as such, as body self-aware, the body has set aside the very limitations of body vis-a-vis mind and has gone on to realize that which characterized these very limitations: cognition, volition, the intentional sphere which contrasted with and delimited body as such (see Chapter II A, contrast of mind and body). The body has passed beyond the limits of the notion of body by incorporating the further richness of being which constituted the restricted nature of the body: that which was not self-aware, which was cognitionally and volitionally opaque, that which in itself fell outside the sphere of intentional life. By incorporating its restrictions body has become corporeal mind. But insofar as mind is the richer perspective, that which limits body and which incorporates the overcoming of these limitations by being self-aware corporeal existence,

[48] *Ibid.*, #88, p. 213.
[49] Hegel, *Werke*, Vol. 4, *Wissenschaft der Logik*, p. 177.

one can term this being: mind incarnate.[50] It is not as if mind were a spiritual thing becoming incarnate via an interaction with a material thing, the body. On the contrary, mind presupposes the body as that which it realizes. The concepts, though, remain distinct, the difference is not conflated. But still the mind constitutes the unity of mind and body by not being an abstraction, but a rich significance ingredient in the meaning of being; it realizes what is implicit in body by being body's self-awareness.

Mind and body contrast with and limit each other. But it is mind considered separately (abstractly) that limits body from its complete truth as body self-aware – it is what mere body is not. Yet body is what mind presupposes for its concreteness, for its being the awareness of reality in general and the self-awareness of a portion of reality in particular. In being for itself, in constituting its own concrete unity, the limitation that is the basis for the contrast of mind and body becomes a moment of the identity in difference of the existing unity mind-body. It must be remembered, after all, that we start with what is a unity in being; the only problem is one for reflective experience, which notes the contrast of mind and body and recognizes it as a contrast to be understood. The contrast is thus finally understood to be real but resolved in the actual dialectic of being and thought. The un-encompassing possesses a partial integrity that is problematic until it is assimilated to that which realizes the un-encompassing as a moment of a richer being, its own, that of mind. The moral is that an understanding of the organicity of being requires that we treat its components not as things but as dimensions or categories of reality which can be accounted for only in terms of the rationality of the concrete: the dialectical nexus that expresses the nisus of thought from less to more encompassing categories of reality. Only then can the relation of two such principles ingredient in being, mind and body, be understood. In philosophy one must grasp the rationale of the real, not describe the interaction of reality.

D. AN ANSWER TO THE *QUID JURIS*

In this chapter we have exhibited the dialectical affinity of categories and isolated two aspects of their relation: the regressive dialectic and the progressive dialectic. These two movements of the dialectic are the grounding of the categories in thought (see Chapter IV A2), they explain *why*

[50] See Hegel's treatment of *Grenze* and *Schranke* in the *Wissenschaft der Logik*, Vol. 4.

the relation of the categories grows out of their significance. The explanation is in terms of reason's goal of explanation: categories *must* be enriched by higher orders of significance as long as the categories are incomplete. The *must* is the dialectical *must* of thought itself – the categorially less complete must be understood in terms of the more complete. Our phenomenology merely exhibited the categorial elements of appearance: mind was related to body via the fact of its presupposition of body. This *de facto* relation of significance has now been understood in terms of the affinity of categories within thought. Their articulation is justified through thought's nisus towards further explanation and therefore more encompassing categories. Their relation is thus *de jure* through legitimation by thought and thought's requirements; the relation answers the *quid juris*, the quest for justification. This represents an advance over Chapter II. The categories of being have been recognized explicitly as articulated and embedded in thought; the otherness of being has to that extent been overcome.

Thought's drive towards richer categories has been understood through positing a final terminus or goal: the complete category, absolute mind, the ground of all significance. If one should find such a heuristic axiom offensive, there is comfort in the fact that the above dialectical conclusions were never directly drawn through its aid. The notion of reason's nisus to greater explanation, not the notion of absolute mind or a final explanatory stance, is the critical postulate. (See Chapter I C2, III B, IV A2, etc.) This is clear from the three principles of the dialectic outlined at the end of IV A2. The notion of absolute mind or a final explanatory stance serves rather to give a full picture of reason's consummate goal and thus a richer explanation of the relation of being and thought. But one can, if one should so wish, accede to the movement of the dialectic without specifying what its final goal would be, though the logic of the situation seems to compel upon us a category of the full explanation of being-in-thought. If this is done, then our discussion of the final stance of explanation can be treated merely as an introductory and now dispensable myth. The allusion to the absolute can be exorcized. What remains then is the central leitmotif: reason is interested in encompassing being and the categories can be ordered in terms of this interest. The final consequences or features of this interest are left unspecified. Yet this is sufficient to elucidate the dialectic of mind and body, once one accepts the legitimacy of reason's drive towards explanation. The full notion of reason's goal provides in addition a perspective from which thought's demands upon being can be understood. It prevents the dialectical cate-

A TRANSCENDENTAL ONTOLOGICAL ACCOUNT   125

gorial analysis from assuming an ad hoc character. The notion of a final explanatory stance then remains necessary if one wishes to specify the goal of reason and thus fully justify employing the dialectic in order to comprehend categorial relations. A stronger reading of the above provides this.

The *de jure* exhibition of the categories gives closure to the phenomenological enterprise via a categorial reconstruction of appearance. In experience we are confronted with a certain given, relatively self-contained reality – a coherent fabric of being-for-us, here that of mind-body. By means of eidetic phenomenology we attempted to analyse out its categorial elements. We were then left with a certain *de facto* structure of ontological factors. By moving from a onesided categorial stance to a more encompassing stance we have reconstructed our original point of departure as a dialectical result. It is a result namely of the above mentioned nisus of reason towards encompassing explanation. That is, examination of the category of body leads to the concrete category of mind (i.e., one in which the category of body is a moment of its significance), which was all along the natural fabric of the appearance with which we were confronted. For that reason one could as well term our progressive dialectic "regressive" – it is a return to the richness of the being with which our investigation began. What we termed the "regressive dialectic" is then a notional account of the initial phenomenological analysis of being for us. This can be taken as an outline of the general procedure of a categorial analysis: i) recognition of a rich and in some respect self-contained reality, ii) phenomenological analysis of this reality into its categorial elements, iii) a transcendental account of these categorial elements – the reconstruction of the original position but this time as a notionally comprehended dialectical result, iv) justification of the result in terms of a general nisus of reason towards complete explanation. The concrete reality of mind-body is comprehended as a complete dialectical result. In this way the dialectical necessity of mind and body, in their identity in difference, certifies the adequacy of the phenomenological description of mind-body given in Chapter II. Of course like all assertions of apodeictic truth by finite minds, the assertion itself is only assertoric. But in any case, the open-ended aggregation of eidetic structures by phenomenology (see discussion of adequacy and completeness of phenomenological descriptions, Chapter I C) is given at least a provisional closure via a system of dialectical necessity.

In fine, this project which began with a phenomenological description of mind-body (i.e., in Chapter II) has been completed by a transcendental turn to the grounding of the categories in thought (i.e., in Chapter IV). Chapter III was the turning point where the ontological requirement of encompassment was explored and yet found to be insufficient for a complete account. Chapter IV completed the account by justifying the necessity of the relation in terms of the dialectical affinity of the categories. This involved transforming the phenomenological (ontological) requirement of encompassment into a transcendental requirement of encompassment, a move from a description of *de facto* structures to a *de jure* account. The result has been the emergence of a transcendental ontology of mind-body: the explanation of the categorial structure of mind-body in terms of its dialectical necessity. What remains for us now is to explore the relation between our transcendental ontological account of mind-body and the accounts given by both social and natural sciences. This affords us with an opportunity to bring our account to bear upon the relation of the positive sciences immediately concerned with the study of mind-body. This will strictly speaking be a *scholium* to our transcendental scientific theorizing. In short, we leave our concerns within the area of pure theory in order to illustrate its implications. Importantly these implications include our answer to the traditional question concerning the causal relations of mind and body.

CHAPTER V

ONTOLOGICAL AND EMPIRICAL STRUCTURES

This final chapter is an attempt to see empirical relations from an ontological point of view. To do this, empirical reality and empirical structures will be translated into variants of ontological structures. It is, in other words, an attempt to grasp the rationality implicit in empirical reality, to apprehend categorially that which is not itself immediately categorial. The suggestion is that observed "correlations" between events in diverse strata of being and the unity of different levels of scientific laws are ultimately to be understood through basic categorial relations. In a sense, we are about to attempt (in programmatic form) a modern dress philosophy of nature and mind, a categorial analysis of the necessary notional structure ingredient in the fabric of empirical reality.[1] But just as truly, this is a special variety of philosophical problem-solving – an attempt to bring metaphysical problems into a context homogeneous to thought and thus amenable to solution in terms of thought. Or since this entire work has been such an attempt, this chapter more specifically endeavors to bring empirical consequences of categorial relations within the scope of ontological thought.

The first four chapters have been concerned with the more general task of understanding the categorial relation of mind and body – that is, *understanding* the relation of mind and body, bringing it to terms with thought, seeing its conceptual fabric. This has involved rejecting the reification of what are in essence conceptual questions. Thus, Chapter III criticized previous accounts of the mind-body relation for their attempting to imagine the articulation of two different sorts of substances. The answer was that there is nothing to imagine but rather a notional

---

[1] We must express our indebtedness to Hegel's notion of the philosophy of nature as "thinking" or "conceptual observation" (Vol. 9, *Naturphilosophie* #246), as an attempt to articulate the different levels of categorial significance present in reality.

relation to understand. Here, though, there is surely plenty to be imagined, visualized. Yet the empirical structures appear to be complicated by notional difficulties usually relegated to the philosophy of science or to a metaphysical limbo. Rather than just freeing the comprehension of a dimension of reality from unnecessary reification, we will here attempt to highlight conceptual structures in what is otherwise an incomprehensible connexus of merely physical relations. This is then an attempt to understand conceptually physical reality and in part to answer some problems in the philosophy of science concerning the relation between neurophysiology and psychology, and the relation between the causal structures each science describes.

This chapter will be divided into four sections. The first will review transcendental ontology's relevance to empirical science. The second section will examine the empirical specification of mind's embodiment. It will deal with both an ontological understanding of the specific features of embodiments as well as adumbrate the general possibilities for the construction of empirical laws describing the physical preconditions for mental activity. The third section will investigate the relation between neurophysiologic and psychoanalytic accounts of man in order not only to resolve apparent conflicts, but also to establish the general categorial significance of these sciences. This will include a short discussion of psychosomatic medicine as an example of an empirical science bridging the categorial levels of mind and body and raising questions concerning the general relations between the laws found in each level. The last section will present a few concluding observations.

A. TRANSCENDENTAL AND EMPIRICAL SCIENCE

In assessing the connection between transcendental ontology and empirical science, one should remember that phenomenological description does not occur *in vacuo*, but begins with the presented appearance of the world. It is in that sense itself empirical. The general structure of the world which it explicates does not have to be applied to the world. Indeed, it is the general structure discovered as integral to the world. Further, the notional relations of the elements of this structure are, as we have mentioned, themselves implicit in being (see inter alia Chapter I C2, III B, and IV A2). Being fully understood is dialectical.[2] In short, transcendental ontology has implications for empirical science simply because

[2] "All that surrounds us can be viewed as an example of the dialectic." Hegel, *Werke*, Vol. 8, *System der Philosophie 1: die Logik* #81, Zusatz 1, p. 192.

eidetic structures are indeed eidetic: they are the significance that is necessarily presupposed for a type of appearance. They constitute the bounds of certain domains of appearance and consequently anticipate a priori the general complexion of the world which empirical science can describe. Further, by delimiting domains of appearance, one delimits possible areas of investigation open to empirical sciences. Insofar as sciences are defined by their content, different sciences are anticipated. Moreover, the dialectical apprehension of the founding-founded relation of categories anticipates relations between the sciences concerned with those domains of appearance. This should not be astonishing; insofar as one determines the sort of world we live in, one determines limits for science.

The relationship of transcendental ontology and empirical science is then the relationship between the notional explanation of being and the full empirical specification of its features. For example, empirical science is required to specify what is to be meant by the requirement of a body; transcendental ontology is required to explain why one should expect to find a physical analogue of mental life whenever there is a mind in a physical world. It gives the rationality of an empirical structure. Consequently, transcendental ontology indicates only very general relationships: i) insofar as there are minds in the world, there will be empirically specifiable bodies of the unique class, embodiments of mind; ii) bodies capable of performing physical functions that are analogues of mental life dialectically imply a mind (see Chapter IV B). In short, given the dialectic of mind and body, bodies should (i.e., should of dialectical necessity) be empirically specifiable as embodiments of minds through determining what is physically required for the embodied life of minds. That is, the embodiment of a mind must in some fashion be the physical analogue of that mind, but how this is to be accomplished depends upon the empirical laws of the physical universe. Conversely, psychical states should be specifiable via the physical states of the body, since the body is the necessary moment which the mind realizes in its life. An empirical project is thus legitimated and outlined notionally. The relation of transcendental ontology and empirical science is thus the relation between grasping the categorial structure of being, and determining the physical fabric of being, between understanding being and displaying its nomological structures.

This chapter consequently is an excursion into the philosophy of nature and mind, not just a series of suggestions for the philosophy of science. Philosophy of science involves a critique of science, it is metascience. In contrast, philosophy of nature and mind treats of the general categories of objects studied by the positive sciences. Since the philosophy of nature and mind is often made aware of certain features of the world through science, it tags after science in a way reminiscent of the philosophy of science. But notionally they are distinct; philosophy of nature and mind (as we shall understand it) is the attempt to delineate the general structure of the categories of nature and mind, and thus impinges at once upon both science and the philosophy of science. Unlike philosophy of science, it does not seek primarily to comprehend the rationality of science. Instead, it seeks to comprehend the rationality of nature and mind. It acts directly and heuristically upon science, suggesting where further unities of appearance might be found. Also, appreciation of the categorial relation can directly affect the philosophy of science through resolving conceptual muddles such as dualism and monism. It is in this fashion that the examination of the categorial relation of mind-body serves as a general schema for the appreciation of the conceptual relations between the sciences of physiology and psychology.[3] It translates general empirical and scientific questions into a categorial context and thus allows them to be treated in a milieu congenial to thought.

B. THE MIND'S EMBODIMENT

Chapter II described in most general terms the structure of the body presupposed by mind. Here we will examine the possibility of further outlining this embodiment, in particular determining if embodiment involves further eidetic subrelationships between mind and particular types of bodily processes and functions.[4] We will explore what has already been isolated, if only in a sketchy fashion. We have found (see Chapter II A6) that the possibility of embodiment includes the possibility of i) physically discriminating and processing stimuli according to their general patterns, and ii) responding to their stimuli in respect of their general patterns. It is because of this that we have referred to the body as a universalizing mechanism. We will now further specify mind's relation

---

[3] For the sake of simplicity, we treat together neurology and neurophysiology, and psychology and psychiatry; further, because of its lack of reference to a psyche, behavioral psychology will be viewed as an adjunct to neurophysiology or at least as a natural science. Psychology for our investigation will be the investigation of the dynamics of the psyche.

[4] Hegel also provides an analysis of particular subrelations of mind's embodiment. See, for example, *Werke*, Vol. 10, *System der Philosophie 3: Philosophie des Geistes*, #401, as well as Vol. 8, *System der Philosophie 1: die Logik* #216-223. Our account draws upon his investigations as offering an example of a move towards a categorial and notional understanding of body and mind.

1. General characteristics of bodies

We know that the interaction between stimulus and response manifested in the body must be complex and intricate because: i) the response to any object requires an assimilation of a complex state of affairs (e.g., the size, distance, mass, and surface properties amenable to available sense modalities, etc.), and ii) operation upon an object requires the initiation of rather complex processes (i.e., initiation of forces of certain magnitudes, in certain directions, etc.). This intricacy is ingredient in the particularization of the body in a spatiotemporal sensible world: any particular perspective for stimulus reception and initiation of responses is but one discriminative or operational adumbration of that world over against an indefinite number of other possible perspectives. Moreover, this one particular perspective must be coordinated with other possible perspectives in order to allow for discrimination and operation in the world; that is, perception and action require orientation, and orientation within the world requires the assimilation of complex relationships. In short, the world has the sense of being a maze of physical complexity in which operation and therefore perception and action is likewise complex. Physical complexity is integral to its notion. Hegel provides us with a description of the character of this state of affairs. "The idea as nature is, in the first place, determined in externality, in infinite individualizations beyond which the unity of form is a mere ideal, in-itself existence, which appears as a goal." [5] To be in the world is thus to be in an existence enmeshed in complexity. Consequently, discrimination and operation in a physical world require a complex physical relationship between the perceiver and actor and this complex world.

Since response and operation in the world involve the interrelation of complex states of affairs (i.e., the relation of operations to objects at certain distances in certain directions, etc.), the coordinated interaction of response and operation is correspondingly complex. The degree of the complexity depends upon the complexity of the physical laws of the world, though a minimum of complexity is ingredient in the very notion of a physical world. Thus as a general proposition the mind presupposes a complex locus of interaction because it is in the world, because as a mundane sentient agent-mind it must be embodied in physical processes of complex interaction with (response to and operation upon) the world

[5] *Werke,* Vol. 9, *System der Philosophie 2: die Naturphilosophie* #252, p. 66.

(see Chapter II A, especially 5, 6, and 7). This precondition entails a physical system of parts of structural intricacy, coherence, reactivity, and intricate interaction sufficient to respond in distinct, complex ways to distinct states of affairs. This a priori specification of the mind's embodiment is not sufficient to describe the body, but it does indicate the general conceptual limiting conditions of embodiment.

Thus, the regressive dialectic gives notional assurance that insofar as there are minds in this world, they will have physically identifiable and specifiable bodies as necessary moments of their concrete existence. That is, these bodies will possess a characteristic complexity and unity. Moreover, this embodiment is differentiated so that certain organ systems are more essential than others and thus possess a different status; the body as the embodiment of mind is differentiated insofar as the mind's mundane functions are. This differentiation is a consequence of i) the nature of this world in which objects are differentiated and particularized spatio-temporally, and ii) the mind's finite realization of its existence such that its functions are expressed in certain parts of the world in certain ways. This is attested to phenomenologically (see Chapter II, especially A5 and 6), as well as by the dialectical necessity for a finite differentiated embodiment of the concrete existence of mind (see Chapter IV B1).

Indeed, even if a mind were purely contemplative, but in a spatio-temporal world, this contemplation would be i) complex temporally, and ii) complex spatially because of its physical presence (see Chapter II A5 and 6). Insofar as it was in the world its processes and life would be physical and present in physical events and processes. Further, the mind's continued mundane existence would be dependent upon the continuance (maintenance) of this physical locus. Thus, the moment of embodiment introduces an implicit element of further complexity into the life of mind. In addition to mind presupposing a complexity in its embodiment corresponding to its own complexity (see Chapter IV B1), the very physical nature of embodiment further complicates the requisite complexity (the degree of complication depending on the empirical laws of the world in question) merely in virtue of the mind having mundane presence.

Embodiment is the moment of externality and thus of rampant complexity. This is not an esoteric insight, but an understanding of what it means for a mind to be in the world. A mind cannot be in the world unless that which is ascribed to its experiencing in the world is also ascribable to it as physical behavior. The alternative to this account is a dualism in which mind is estranged from its body. The unfeasibility of that position has already been discussed (see Chapter III A1). We are left then with a conceptually recognized empirical complexity as part of the existence of mind.

This has another side as well, understood in terms of the progressive dialectic. Not only can one discover the physical correlates (i.e., the embodiments) of mental mundane life, but one can also identify the psychical correlates (i.e., fuller significance) of physical processes. This is assured because the category of finite mind is the fuller significance of processes that would otherwise have the truncated significance of being merely physical analogues of mental operations (see Chapter IV B2). In short, the general conceptual relation is the basis of particular conceptual relationships, of particular physical phenomena betokening particular psychological events as integral to their full significance. Combined with the regressive dialectic, it affords a connexus of significance that forms the notional fabric of mind-body. This allows us to make two general statements about empirical relations. i) When there is a mind of X complexity and Y integration performing mundane actions a, b, ... n, then there must be a physical system of X' complexity and Y' integration performing physical operations a', b' ... n'. And ii) when there is a physical system of X' complexity and Y' integration performing physical operations a', b' ... n', then it must be the embodiment of a mind that is of X complexity and Y integration performing actions a, b ... n. For example, i) when there is memory of past events of a certain detail, there must be a data storage system, etc., of adequate complexity to account physically for this. ii) And when there is a data storage system that self-reflexively monitors itself so as to respond to universal patterns, etc., with the same complexity and unity as do minds, then it must itself be a mind with a memory. The further development of these generalizations requires the empirical identification of the characteristics of physical operations that embody mind, and of mental actions that complete the significance of the body.

This identification must be made not merely on the basis of i) temporal correlations but also because of ii) the correlated physical processes' ability to perform functions analogous to mental operations (e.g., physical processes of discrimination of stimuli as the analogue of the perception of objects). Such identification of the empirical characteristics of the physical embodiment of mind could be used to identify the presence of minds in problematic contexts (e.g., computer "brains"). If a physical system performs the physical function presupposed in the embodiment of mind with the same unity and integration as is found in actual embodiments of mind, then it *should* be a mind, and should be treated as such. Moreover, the degree to which this physical system realizes this full function could be used to decide the level of mind present. For example, such criteria could be invoked to decide the level of consciousness present in lower organisms.

Further, transcendental ontology can go beyond description of just the relation of the mind and the nervous system and examine the relation between the mind and all the body's organ systems. It can give certain general limits for embodiment. For example, the body must i) have physical processes of discrimination, ii) structures for the effecting of physical operations, and iii) structures and processes of physical self-maintenance (even if this only includes acquisition of energy). These limits represent certain conceptual boundaries discovered by fictively varying the physical structure of the body to determine what structure is eidetic to it. Moreover, these limits represent a conceptual fabric of notional necessity forming a categorial connexus, the logic of the structure of being. They are further established by showing themselves to be ingredient in the category of embodied mind (a category discovered in reality) as well as being themselves available in reality and moreover betokening the further and richer categorial structure (see Chapter IV). Their cogency is their illumination of the rational structure of being.

Thus, when one examines the significance of the nervous system, one finds that the absence of a nervous system is an absence of the core of mind's embodiment in the world. A system of physical processes such as the nervous system is the uniquely proper material moment of the mind's concrete existence (see Chapter II A and IV B1). Thus Wilfred Sellars's "core person" has an ontological status: it is the primary structure or moment of embodiment.[6] It is the embodiment of the imagining, thinking, and willing of man, and, if further extended, of his perceiving. Without this physical structure mental processes would lack any actual point of contact with the world.

The life support systems of man (digestive, respiratory, etc., systems) are a lower categorial level of embodiment, because they are not the embodiment of a specific feature of mind, but of its more abstract note of life. We use "life" here with specific intent. It designates only the first gleam of mentality: the non-conscious goal-directedness of the physical processes that maintain the unique unity of certain organizations of matter. This note of "life" is presented in isolation, for example, in the case

[6] The endocrine system performs cruder functions of effecting changes, and responding to internal changes, in a manner like the nervous system. As such it is integral to the lower functions of the nervous system both afferent and efferent.

of plants and fungi, which live, but without being an identity pole of active intentions (i.e., those engaged in by a mind).[7] The goal directedness present would be devoid of an active consciousness of an object: no one has directed himself towards these goals, though the goal-directed behavior is unified about an identity pole which is the sense of the permanence of a particular living being. This cannot be considered a true subject pole until it becomes or unless it is the lowest moment of a mind's intentionality. At that juncture the apparently purposeful behavior becomes mental in becoming someone's drives and needs; it becomes a lower moment of a mind's life.

This lower level of existence is presented in involuntary physiological processes that are not just habitual but in whose non-physical significance conscious awareness cannot actively live. For example, one can know of his heartbeat but need not attend to or even ever be conscious of it. It is rather a part of the unity necessary for conscious mundane life, but not itself a part of consciousness. The processes of self-maintenance represent the moment of the mind's embodiment of its needs of maintenance in the world. While these processes are a moment of mind's existence in the world, they can be categorially distinguished from those physiological processes that are the embodiments of imaginings, thinkings, and perceivings. They are a moment of the physical duration of the higher systems, abstracted from the richer significance of embodying particular mental functions. They are the physical reality of perdurance in the world apart from which the body would either be causally isolated from the world or devoid of energy if not of structure.

The musculoskeletal system, on the other hand, is a moment of a further realization of mind – practical mind.[8] It is the embodiment of

---

[7] Unfortunately, this reveals a question that can only be alluded to and cannot be considered at length. Namely, there is a sense in which a living being is presented as a special unity apart from any question of considering its status as a mind. It is recognized as an object organized as a pole of teleological behavior. The goals of the organism, though they transcend the stratum of physical meaning, are not necessarily conscious intentions. They may, though, always have the structure of a fabric of feelings. In that case, life would always be coterminous with mind. It would be a categorially distinct but notionally degenerate level of mind. Or on the other hand, life may be only a new level of unity within the domain of physical reality, but distinct from *merely* physical reality in that it would have a unity that allowed it to appear "as if" a mind were present. That is, living things appear as if they were oriented towards goals (e.g., their own preservation). This question is suspended (as mentioned in Chapter II A7) since we are dealing with the case of human life, where lower moments of embodiment can be treated as special moments of mental life. This leaves open the questions of the relation between conscious feeling, and organic teleology.

[8] Hegel, for example, refers to the efferent nervous system as the practical nervous system. *Werke,* Vol. 9, *Naturphilosophie* #354, Zusatz, p. 594.

mind, no longer as a mere spectator of objects but as the producer of objects (see Chapter II A5 and 6). In its absence, there is the absence of mundane agency as well as the absence of a richer significance for mind. As the embodiment of practical mind, the musculoskeletal system brings the nervous system into a richer domain of significance. That is, practical mind as the effector of an objective order of reason embraces theoretical mind as a moment of its existence in the world. It is the physical reality of actualized mundane volitions.

Consequently, there is a categorial distinction of three classes of organ systems according to their status as necessary moments of mind. These relationships allow a notional understanding of organ systems in terms of the general requirements of embodiment: i) nervous system – the embodiment of thought and perception; ii) the neuromusculoskeletal system – the embodiment of action; iii) the respiratory, digestive, excretory, and cardiovascular systems – the embodiments of life's persistence.[9] The unity of these systems is the unity of mind; they are its distinguishable moments. Their distinction is the basis for a conceptual understanding of the structure of embodiment, for a general outline of the relations between mental life and physiological processes. We will later see to what extent this solves problems.

2. Organ replacements

The above distinction of levels of embodiment and the concept of a dialectic of mind and body allow a notional explanation of the integration of foreign organs into the life of a body and thus of a mind. The body is mine in a fashion unique among the possible relations between mind and physical objects.[10] This singular relation, which is encountered in appearance,[11] has, as we have shown, a dialectical basis. On the one hand the mind requires an organized portion of matter that can perform the physical analogues of mental life. The requirement is not for parti-

---

[9] In this sense it provides the necessary categorial machinery to account for the possibility of knowing other minds. In that only the bodies of others are given, knowledge of others remains very problematic unless a categorial account of further levels of significance is available. Once available the actual everyday experience of others can be accounted for in terms of the higher levels of appearance that are dialectically related to and complete the lower levels and thus are presented with them.

[10] In this analysis we will employ the significance of "mineness" which was discussed in Chapters II A and IV C.

[11] See Chapter II B; also, Husserl has given a perceptive analysis of the constitution of the significance of the body – see *Husserliana*, Vol. 4, *Ideen II*, especially #35-42.

cular parts or arrangements of parts but for a certain level of operational complexity and unity. The body and its parts are peculiarly "mine" in that they, out of all other material structures, carry out the physical reality of my life. Organs receive their status as "mine" through performing a role in a particular system of life and action. In being integrated into this structure they enter into the dialectical relation of presupposition and fulfillment which characterizes mind-body. On the other hand, the body requires the mind for its full significance. The body is engaged in certain activities by means of internal processes with a level of organization which is not fully understood save as also being the organization of mental activity (see Chapter IV B2). Thus neither consciousness nor even its incarnation is a thing. They are rather intricate activities related to each other in virtue of the significance of their organization and function. The dialectic is after all a relation between the complexity and unity of mental and physical processes. And function is reality seen as a signal act of being significance. It is being in a process of meaning, often emerging into new levels of existence.

Insofar as the mind-body relation is a relation between levels of function, the distinction of organ systems for maintenance, discrimination, and operation differentiates levels of complexity incident to integrating organs in these systems. The crucial stage is the incarnation of thought (the nervous system). The mere embodiment of life (the maintenance system) appears as a preliminary stage and the embodiment of action (the ennervated musculoskeletal system) as a further development of the embodiment of thought. Thus replacement of organs associated with life support seems least problematic and replacements of portions of the central nervous system most problematic, since the latter are concerned with the very core of mind's embodiment and involve a unique level of material organization. With the latter one must imagine minute and gradual replacements. That is, the problem of physical continuity is most crucial since mental continuity is so closely dependent upon it. The replacement of limbs is again less problematic, for as mentioned it already presupposes the full incarnation of mind as perceptive and cognitive.

In the case of organs concerned with life support (e.g., the heart, lungs, kidney, etc.), their mineness, and thus their status as integral to the embodiment of a mind, is realized in their proper performance of fairly discrete functions. They can therefore be easily replaced by mechanical organs which would become "mine" insofar as they reproduce the functions of a normal organ. This is perhaps most easily seen in the case of minor replacements such as heart valves, where the irrelevance of the cellular substructure is most evident. The cellular substructure is not integral to the valves' regulation of blood flow. Their function in the body's economy is unaffected by their synthetic origin (except insofar as synthetic valves are not yet as mechanically reliable as are "organic" valves). Rather, their organic role is determined through their performance in processes with particular levels of complexity and unity. Thus, the turn from metaphysical ontology to categorial ontology causes one either to abandon the old quandaries about inorganic components of living structures or relegate them to empirical science. The philosophical assessment of the inessentiality of cellular structure opens the possibility for a metaphysically simpler situation for the embodiment of life and mind. If a cellular substructure is required for life, this is a purely empirical necessity.[12]

Organs embodying higher categories of significance provide the most interesting cases of transplantations. In a limb there is the embodiment of both perception and action. One must feel (including kinesthetic sensation) and act with a limb in order for it to be integral to his body. The criterion for *my* limb is not physiological history. Rather, the limb's sense of mineness lies in its executing the psychical and thus physical processes that are integral to feeling and acting. The operations of the body must flow from thought and its physical embodiment if these operations are to be actions, not merely events or miracles. That is, there must be a unity of volition and operation in order for a limb to become the organ of one's action and perception. Phenomenologically, the constitution of "mineness" rests upon the automatic "if-then" relationships between kinesthetic flow patterns of sensations and other correlated sensations (touch, etc.) that become objectified as contacts with the world.[13] Or more ontologically put, when a prosthetic limb can discriminate objects in a precise manner and execute intricate movements proper to a limb (all in relation to the discriminations and movements of the body as a whole and as a consequence of central nervous initiation) and thus allow me to sense fully and act through it, then it would necessarily be *my* limb. This is not primarily a prediction of a psychological state but a descrip-

[12] Indeed, one can ask further whether a description of the body as the embodiment of mind must extend beyond a description of cells, molecular structures, and energy transfers, to a description of the part played by particular atoms. This is an empirical and perhaps even a metaphysical question, which can be answered only by inspecting the particular details of the case. It is a question of whether and to what extent microstates are specifically integrated into and specifically characterize the macrostates that have been discovered as embodying mind. In any case, it exceeds the scope of our investigation.

[13] For a description of the role of these "if-then" relations, see *Husserliana*, Vol. 4, *Ideen II* #36-40, especially #40.

tion of an ontological stratum delimited by the significance of feeling and acting in a physical object. Just as with one's natural embodiment, it is another question as to when and how perfectly a prosthesis is integrated into one's body image.[14] Indeed, a certain lack of integration would be merely a deficient mode of "mineness" as long as there was sensory and motor function through the prosthesis. The necessary degree of integration for sensory and motor functions is an object for empirical investigation.

The possibilities of replacement can be carried even further; one can envisage replacing parts of the brain damaged through cerebral vascular accidents, etc. In fact, the brain may be replaceable piece by piece, until the original material is gone but the function and significance remain unchanged. This would not be much different from the normal replacement of material by the body. As long as the replacement was gradual, allowing replacements to be integrated, the brain's "person" would perdure. In short, the mind is not a thing but a process [15] of thinking, willing, etc., a stratum of ontological significance that presupposes a physical embodiment of these mental activities. There is no reason to assume that this could not be accomplished by a mechanical brain. But the actual determination and specification of this possibility is dependent upon empirical laws specifying the requirements of embodiment.

C. STRUCTURAL INTEGRATION AND INDEPENDENCE OF MIND AND BODY

1. *The substructure of consciousness*

The physical processes that embody mental activity are not merely signs or indices of the presence of a mind. They are the presence of mind in a

---

[14] The body image is the conscious and unconscious realization of the cerebral representation of all one's somatic sensations as organized in the parietal cortex – a concept with obvious categorial depth. See, for example, *Dorland's Medical Dictionary*, p. 725: "A three-dimensional concept of one's self, recorded in the cortex by the perception of everchanging postures of the body, and constantly changing with them." The body image is influenced by numerous factors: past experiences, mental health, social environment, etc. Moreover, it changes slowly as attested to by the fact that 98% of limb amputees experience the phantom limb phenomenon (Lawrence C. Kolb, *The Painful Phantom* [Springfield, Ill: 1954].). Thus, difficulties would be expected in the integration of a new organ into one's body ego or sense of physical presence in the world.

[15] In contrast to Kant, Hegel affirms that mind can be an object of investigation without becoming a thing. See *Werke*, Vol. 8, *System der Philosophie 1: die Logik* #47; Vol. 19, *Geschichte der Philosophie 3*, pp. 577-579. Hegel characterizes this reality of mind as a process of self-realization. "Only if we view the mind [*Geist*] in the described process of the self-development of its notion, can we apprehend it in its actuality." *Werke*, Vol. 10, *Philosophie des Geistes*, #379, Zusatz, p. 17.

physical world; this is encountered on a macro-level in the observation of the anger, happiness, etc., of one's fellow men. Their mental life is presented in the behavior of their bodies. Happiness is embodied in, not merely inferred from, a smile. But a further awareness of the structure of this embodiment does not lead primarily to an examination of the mechanism by which the risorius muscle et alia form the physiognomy of a smile. Rather, it carries us to the neurological basis of the smile; one can be happy even if one's facial muscles are paralyzed. This level of physiological investigation is not merely physically more basic than an observation of gross behavior. It has the categorial status of being the explication of the specific embodiment of mind (see Chapter IV B1).

But there are special difficulties here. In that one recognizes neurophysiological processes as basic, one is confronted with the question of their causal determination of conscious mental life. As we will see, one is also confronted with the question of the influence of subconscious processes upon conscious mental life. These questions arise out of the complexity of embodiment and are themselves interwoven. For example, the possibility of subconscious mental life broadens the notion of embodiment. It forces us to recognize a greater complexity in the mind-body relation. In such a state of affairs, embodiment would include various degrees or kinds of consciousness embodied. An example of such embodiment would be something like this. If a person steps on a tack (all other things being equal, i.e., one is not under local anesthesia, suffering from a neurological disorder, etc.), a free nerve ending in the dorsum of the foot is stimulated, causing it to discharge its action potential. There may be some sense in saying that the cell feels pain (see Charles Hartshorne, *The Logic of Perfection*, Chapter 7). The potential is propagated up the dendride of a dorsal root ganglion and from thence up the axon into the substantia gelatinosa of the lumbar section of the spinal cord. There a synapse is effected with the dendrite of some proper sensory neuron whose axon crosses to the contralateral side of the spinal cord. The impulse is modified at this level as well as there being the initiation of a withdrawal reflex. This is the beginning of gross pain behavior, even though there has not yet been a conscious awareness of the pain. One may, though, want to say, in some perhaps metaphorical fashion, that there is at this juncture a spinal "awareness" of the pain.

The impulse then proceeds up the lateral spinothalamic tract into the ventral nucleus of the thalamus. At this point, great modification of the stimulus occurs, especially by the reticular activating formation. It is here that some subconscious awareness of pain may have its embodiment; they, if they are anything, must be some general structurings of the pain. Finally, at this level the subject may also first begin to experience pain (i.e., a thalamic awareness of pain). From the thalamus the impulse can pass into the gyrus postcentralis of the cerebrum and be referred to the portion along the longitudinal fissure. At this point the subject will experience full conscious pain. But the pain is already being altered by past experiences of pain, that is, by past conditioning which is stored as "memories" in the temporal lobes and as certain reaction patterns in the frontal lobes and elsewhere. Even after the first experience, numerous modifications, conscious and otherwise, will occur, inhibiting or accentuating the pain impulse. With these modifications the subject will experience different variations in his sensations. In this, not only will the brain as a whole be involved, but perhaps various subconscious levels of experience will be unified in the full conscious experience of pain. Moreover, the unity of this full conscious experience will embrace these physical processes and their subconscious realizations as moments of mind's embodiment and concreteness. They will not constitute separate elements, but distinguishable moments completely understood only in terms of self-conscious mind.

We are led then to the problem of subconscious mental life, not merely as the realization of neurophysiological processes but as a component of mental life. That is, in accounting for the fabric of conscious life, one is often led to posit grounding levels of mental activity into which self-consciousness cannot directly enter. These structures (if in fact they exist) are still a part of the structure of mind; they are components that remain merely on the level of feeling or subconscious intention without becoming immediately the content of a self-consciousness. They enter self-consciousness only mediately by providing content and a substructure for higher strata of mental unity. Moreover, such processes need not be posited merely as empirical theoretical constructs, as they are at least in part by Freud. They can also be sought as notionally and phenomenologically ingredient in man. For example, if Husserl is correct, "subconscious" strata of the life of mind (such as the automatic protection-retention structure of inner time consciousness) are eidetic structures integral to consciousness. Further, subconscious processes must at least in some fashion contribute to the temporal continuity of mental life, filling in the expanses between conscious attention with a fabric of subconscious intentions. Otherwise, mental life would in a certain sense appear to be a periodic phenomenon, disappearing in deep sleep. In such a state of affairs, mental life would be dependent upon the body not only via the founding-founded dialectic but because the mind would lack any continuity of its own. That is, because of mind's categorial independence one would be tempted to search for an existential independence and continuity. One might answer, though, that the continuity of mental life is not established in objective time but through the mental structure of protentions and retentions forming the continuity of subjective time. One does not normally experience actual lacunae in his mental life. This continuity, though, is probably indeed psychologically dependent upon preconscious mental structures.

In any case, since subconscious processes are to be understood as part of the structure of mind, then the question of their determining consciousness arises. But subconscious levels of mind can not fully determine consciousness without excluding the possibility of consciousness; if consciousness is distinguished from the subconscious, it must possess its own intrinsic structures. For example, eidetic to full self-consciousness is one's recognition of his mental activities as his activities. In this an element of categorial novelty emerges. Though the subconscious processes may produce material for consciousness, they cannot determine the recognition itself. At least insofar as one's approval can be seen to follow from self-consciously desired, rational principles, there is velleity if not volition: the volition lies in the conscious nisus towards a state of affairs, a nisus which rationally follows from one's principles or even desires. Recognition of the approval as approval of one of his desires requires at least a rational consonance between the approval and one's own desires or principles. That is, one must understand the consonance, it must not be a brute fact. Such an approval is then a categorial novum transcending any subconscious processes and thus describable according to its own laws. Volitions follow from rational principles which are then the *reasons* for, not just the causes of, the volitions. As long as the causes of actions can also be understood as reasons and as long as these reasons "fit in" to one's patterns of rational action (i.e., the reason is consonant with the logic of rational action), then these are volitions even if the subconscious substratum of mental reality is bound together by rigid "deterministic" laws.[16]

What is decisive is that volitions be materially and formally a part of a system determined according to indigenous, rational rules. They must belong materially, they must appear as elements of rational conduct, and

---

[16] This characterization may be too narrow and exclude capricious volitions. To clarify such questions would be to enter a tangential but separate investigation. It is enough if a successful, though restricted, account of the compatibility of causes and reasons is indicated.

they must belong formally, they must appear articulated in rational conduct. Once these criteria are satisfied, volitions possess a reality independent of any causal substrata and it would be a category mistake to picture a collision. Indeed, to picture such a collision would be to reify different categorial levels of significance and then wonder how different substances or substantial causes can be interrelated: a repetition of the dualist-monist fallacy. Instead, through distinguishing categorial levels without trying to impose a pictorial schema of interaction a new level of meaning can be understood to be present and independent just as it is indeed apprehended. One apprehends himself as free without having to renounce the strictness of underlying causality; the physical substratum appears rather as the substructure for something more.[17] This is perhaps clearer in terms of the relation of physiological and psychological laws.

The physical moment of existence is a structure of reality invaginated in the higher and richer stratum of mental significance. The mental stratum embraces the nomological structures of the body within a new structure of laws so that: i) though every mental nomological structure has its physical analogue or embodiment, ii) the mental stratum goes beyond the significance of the physical stratum, and iii) has a necessary structure of its own. For example, the structures of inner time consciousness are binding as such, even though they presuppose that the body is engaged in a physical process of "information" retention and "anticipation" that coordinates any moment of stimulus acquisition. That is, time consciousness cannot exist without retention of past "nows" and implicit protention of further "nows"; without this structure it "collapses" into an eternal now. Such a nomological structure (i.e., "any 'now' is found in a structure of retended and protended 'nows' ") is a purely mental law: it is understood without reference to physical nomological structure and is binding in its own right.[18]

Of course, such an understanding is abstract, for one must in addition note the physical foundation of this psychical nomological structure. But it does mean that the laws of mental life have a structural independence. This would be the case not only insofar as a law expressed an eidetic necessity (e.g., such as Husserl asserts in *Ideen II*, #63-64), but also insofar as it was understandable purely in mental terms. Thus, explanation

---

[17] Kant at least in part saw this point. "I assert that every being that cannot act save according to the *idea of freedom* is for that very reason, from the practical point of view, actually free. That is, all the laws that are integral to freedom apply to that being just as if its will were shown to be free in itself by a valid demonstration in theoretical philosophy." *Werke IV, Grundlegung der Metaphysik der Sitten*, p. 448.

[18] This problem is discussed by Husserl in *Husserliana*, Vol. 4, *Ideen II*, #63.

of action in terms of drives, desires, etc., has a significance apart from the physiological substratum of those mental processes. The independence is based upon the conceptual integration of mental processes, the essential distinction of mind and body (see Chapter II). Insofar as mind is distinct from body it has a distinct structure that remains independent of the structure of the body even if the mind presupposes that very structure. In short, if an intention is not a physiological process, then neither is the psychical structure of intention a physical structure of physiological processes. They are distinct logically and phenomenologically and thus ontologically.

As a consequence, one must understand the relation of the psychical and physical strata of meaning as having this organization: i) each stratum possesses its own nomological structure with its own necessity. ii) The embodiment of mind requires that the mind reflect the physical structure of its body by possessing the structure as a concrete moment of its own significance. The mind is determined by the body in being dependent upon its functioning. The mind can perceive and act only insofar as the body can receive stimuli and operate upon the world. The body thus sets limits and determines the patterns of functions within essential limits (see Chapter II A and IV B1). iii) But this does not imply that the structure of mind is therefore fully determined by the structure of its body. On the contrary, there are, as we have mentioned, psychical nomological structures into which the physical structures must "fit" *if* the mind is to be embodied. Thus, one can as easily maintain that the physical structures of the body are determined by the mind; that is, insofar as a body is to be the embodiment of mind it must possess structures consonant with those of a mind – i.e., physically allowing the embodiment of mind (it must be the protasis in a progressive dialectic, see Chapter IV B2). iv) Thus, both mind and body are co-determining, since a) mind is the higher truth of the body and that to which the body must conform if its significance is to be fulfilled (e.g., an alternative is the death of the body), and b) the body is the necessary moment of the mind and that to which the mind must conform if its significance is to be concretely present in the world.

The relation between these two strata is thus fully dialectical: the physical stratum is enriched within a stratum of mental structure which is compatible with and completes the physical stratum. The search for a causality bridging mind and body is not only superfluous, but a mistake in categorial analysis. The strata appear interrelated categorially, not woven together in a causal nexus. Moreover, by the nature of this relation any actual efficient mental cause would of necessity have its physical moment and thus physical efficacy (i.e., a necessary moment of its reality, see Chapter II, IV B1 and the foregoing paragraphs). Thus a head-on conflict between two different sorts of causality is precluded. All efficient causality is at least in part physical. The *mental* significance of the efficacy would lie in the cause's role in a level of meaning that completed but did not exist apart from certain physical organizations and their processes. It would not be as if "the body" determined a second thing "the mind," or that "the mind" determined a second thing, "the body." It would not merely be that the interaction was instantaneous or parallel. On the contrary, the relation must be one of significance, not causality or force. The chain of "mental causality" is not a line of causality separable from the physical chain, rather it is its categorial completion. In any case, a two causal-chain model would only be the re-emergence of a dualism as well as a denial of the presented structure of appearance (see Chapter II). Indeed, mind-body has a dynamic unity that allows one to distinguish distinct levels of nomological and causal patterns while recognizing that these levels must coalesce as one reality. It is causally and substantially one in its several distinct strata of significance.

This promises the possibility of a doctrine of integrated causality and nomological structure so that one can recognize distinct levels of significance as well as unity of efficacy. As mentioned, this unity would include subconscious drives. They can have their efficacy only through their physiological substratum, while their nomological place is fully determined by laws of mental structure, albeit subconscious structure. A subconscious mental life would form as it were physiological reality's first step into the life of mind – a step that would determine higher levels of reality not by efficient causality but by providing certain necessary moments of mental reality. In short, the same categorial relation would obtain between subconscious processes and physical processes as between mind and body in general.

In conclusion, we find that conscious life has its own character and integrity including the possibility of freedom. Because of the identity in difference of mind and body, physiological processes can be recognized as following their own rules without undermining the autonomy of mental life. One may speculate whether physical indeterminacy is associated with mental life. If there were a great deal of causal indeterminacy ingredient in neurophysiological processes (which offhand does not seem to be an unlikely hypothesis), then elements of mental life such as capricious actions would have a welcome, pictorially appropriate substructure. On the other hand, one should not be astonished if rational behavior is "reducible" to rather determined patterns of neurophysiological behavior. Reduction in a proper sense requires a distinction between adequacy and encompassment (see Chapter III A4). Physical efficient causality is all inclusive, it touches all entities in the world but it is not all-encompassing, it does not determine all dimensions of reality. The note of rationality and freedom is after all not under any circumstances to be found on the level of neurophysiological processes. It would, though, have to be the case that only certain patterns of neurophysiological activity would be sufficiently complex and unified in order to be the substructure of rationality. In any case, neurophysiological processes, as well as subconscious mental life, underlie but by no means preclude, indeed rather found, the full richness of human life.

This can all be summed up in a dialectical principle: a higher category of being further determines a lower nomological structure by completing it in a new dimension of reality that has its own laws, but laws that are laws of a further dimension and thus that always refer back to the previous dimension. Each level has its own integrity, but they are integrated not causally but in virtue of a progression of meaning. A bit poetically, the body predestines the possibilities of mind – its physical possibilities; mind, though, is the destiny of body, the completion of its significance. The matrix of physical reality with its causality and laws provides the basis for mental reality, which through its own unity and nomological structure adds a new ontological depth to matter. A mind is not immediately, causally determined by its body – its body is its physical context. A body is not immediately, causally determined by the mind it embodies – the mind is the body's full meaning, realized in a new dimension of reality.

2. *The rivalry of teleology and mechanism: an etiological dualism*

The categorial relation, which grounds the relation of physiology and psychology, if understood, obviates a problem symptomatic of latent dualism: the rivalry of teleological [19] and mechanistic explanations of human behavior. This understanding gives as it were a categorial diagnosis of a pseudoproblem. It resolves the mind-body quandary at the level of the unity of the sciences. For example, Norman Malcolm alleges that an all-embracing neurophysiological account of human behavior precludes teleological explanation.[20] He asserts that "no nonteleological

[19] Teleology is here used to indicate intentional teleology: that is, that a mind is directed towards a *telos*.
[20] "Explaining Behavior," *The Philosophical Review* 76 (January, 1967), pp. 97-104, and "The Conceivability of Mechanism," *op. cit.*, 77 (January, 1968), pp. 45-72.

sufficient causal explanation of the behavior that is correctly explained in terms of purposive principles could turn out to be true. There are limits to what empirical science could establish." [21] He presupposes that in choosing between teleological and nonteleological systems of explanation one is choosing between two exclusive types of causality. This is based on a false premise concerning the categorial relation between strata of significance; "to find a sufficient nonteleological causal explanation of all our daily behavior would be to find that people do not have desires, fears, or goals." [22] Such is a reifying judgment about the structure of the world. It assumes that either mental phenomena form causally efficacious plugs in special gaps in the fabric of human behavior, or they are nothing at all. This is a category mistake of placing thoughts and things in the same stratum of significance and then trying to see how they could possibly fit together. The result is metaphysical rivalry between types of causal significance: final and efficient.

The rivalry between psychological and physiological explanations rests upon a confusion, a fear that adequacy precludes encompassment. That is, an explanation including all entities in the universe is seen as challenging an explanation that would encompass a further dimension of reality possessed by only a restricted class of entities. This is a re-emergence of the dualist-monist controversy. Physics' claim of an adequate explanation is taken to require that either psychology be reducible to physics or the claim of physics be challenged. The quandary results from attempting a unidimensional answer to a categorial complex problem. The answer lies in recognizing that psychology's claim to encompass categorial unique realities does not conflict with the claims of physics (see the discussion of adequacy and encompassment in Chapter III A4). It involves psychology examining a nomological superstructure that completes but does not negate a physical substructure, the province of physics.

Norman Malcolm's interpretation of the challenge of neurophysiology exemplifies the dualist reading of the relation of mechanism and teleology. It, though, represents a failure to recognize certain important features of the structure of mind-body that are worth reviewing. i) As we have just said, a neurophysiological account can be all-inclusive at its level without excluding the possibility of a teleological account (see Chapter II B). ii) Indeed, neurophysiological causality is only fully understood in terms of the richer causality of mind. That is, the presence of such an

[21] "Explaining Behavior," p. 104.
[22] *Ibid.*, p. 103.

all-embracing fabric of physical causality does not preclude, but rather requires, a further stratum of teleological causality that would include the behavior of organisms and especially certain organisms. Further, such a stratum of purposive explanation presupposes the physical stratum (see Chapters II and IV B1) for its real presence in the world. Thus, far from denying the efficacy of purposive action (of purposes and intentions molding the world), the stratum of physical causality is its necessary moment. iii) Nor does this mean that actions and purposes are reduced to physical occurrences. Quite the reverse. Neurophysiological processes are fully understood only in terms of their being a moment of mind's reality. The causal efficacy of minds is by no means challenged because actions constitute the full causal being of physical causes (see Chapters II B7, IV B2 and V C1). iv) Thus the complexity of the causality associated with sentient beings must be recognized as a dialectical founding-founded structure of physical and psychical significance, which is a single determining unity. It does not determine things twice over (as would be the case with two parallel redundant chains of causality), but once fully. The paradox of the opposition of neurological and teleological causal determination is solved by successfully analysing the unity in the complexity of human causality and human reality. Thus, it is not the case that "a comprehensive neurophysiological theory would leave no room for desires and intentions as causal factors." [23]

On the contrary, the dialectical relation of mind and body insures the possibility of an investigation of the whole man; a final choice between either a complete neurophysiological explanation of man or of a purely psychological study of his intentions and actions is impossible. These studies are distinguishable, but not separable, save abstractly. In reality, they are integrally related in that their objects are aspects of one dialectical whole: man. As in the ballet, there cannot be a separation between the physics of man in motion and the cultural significance of his acts of movement, so in an encompassing study of man the physical reality of man must be treated as integral to the richer significance of man as conscious actor. We are left then with the categorial basis for a general program of articulating the organic unity of man and of the disciplines that investigate him.

D. PSYCHE AND SOMA

The guiding role of the categorial relation can be illustrated in the analysis of the etiology of diseases with pronounced physical and psychical

[23] Malcolm, "The Conceivability of Mechanism," p. 63.

expressions. Such diseases can occasion oppugnant accounts, one physiopathic and the other psychogenetic. It is for such recrudescence of dualism that categorial analysis can be of etiopathic relief.

1. Psychosomatic medicine, psychoanalysis, and psychophysiology [24]

Ulcerative colitis is a disease that contributes an example of categorial amphibiousness. It is defined as an "inflammatory disease involving primarily the mucosa and submucosa of the colon. The disease is peculiar to man, does not appear in epidemic form, is not contagious, and relapses frequently. In spite of long periods of remission, the likelihood of recurrence is ever present." [25] Though there are two distinct accounts of its etiology, the information supporting either side is still indecisive; it remains in the epistemic limbo of idiopathic diseases. On the one hand the pathogenesis is interpreted as purely somatic. For example, Zetzel finds that the disease is very probably a variety of autoimmune or hypersensitivity reaction.[26] On the other hand, purely psychoanalytic explanations of the genesis of the disease have been proffered. "Ulcerative colitis is either initiated or may exacerbate in some patients as a reaction to any form of extreme stress. The attack may commence three or four weeks after an unforeseen threat to security in the form of bereavement through death, separation from a significant person, rejection, disillusionment, the loss of a part of the body, or a change in psychologic status with

---

[24] These terms are often given quite restrictive definitions. Here we should use them in a broader sense. Psychosomatic medicine (disorders etc.) will be used to indicate the larger themes suggested by psychosomatic – "pertaining to the mind-body relationship; having bodily symptoms of a psychic, emotional or mental origin." *Dorland's Medical Dictionary*, p. 1245.
This is especially true with regard to psychoanalysis, which far from being understood broadly as the analysis of the psyche, is often taken to refer to the orthodox Freudian method of psychotherapy – "the method of eliciting from patients an idea of their past emotional experiences and the facts of their mental life, in order to discover the mechanism by which a pathologic mental state has been produced, and to furnish hints for psychotherapeutic procedures." *Dorland's Medical Dictionary*, p. 1243. "Traditionally, classical psychoanalysis has referred primarily to Freud's libido and instinct theories; recently, it has come to include the concepts of ego psychology as well. Essentially, it is based on the free association method of investigation which yielded the data used by Freud to formulate the key concepts of unconscious motivation, conflict, and symbolism which formed the basis for his broader theoretical system." A. M. Freedman and H. I. Kaplan, ed., *Comprehensive Textbook of Psychiatry* (Baltimore, 1967), p. 269. The term psychoanalysis has been broadened by subsequent workers and comes now to include such theories as "existential psychoanalysis." We will employ it broadly to indicate an analysis of conscious life in terms of underlying psychical forces, usually with a view towards therapy of disorders in mental life. See also footnote 39.

[25] Louis Zetzel, "Ulcerative Colitis," in *Cecil-Loeb Textbook of Medicine*, ed. Paul Beeson and Walsh McDermott (Philadelphia, 1967), p. 924.

[26] See *ibid.*, pp. 925-927.

consequences of diminished self-esteem... Thus an important factor in management is the establishment of a solid dependency relation whereby the patient can gain a feeling of mastery over himself and his environment. Dramatic interruptions have been obtained when the physician has assumed a firm protective role and has given special attention to the patient's need for support." [27] Further, the psychogenesis of the disease is recognized by the American Psychiatric Association in that the disease is officially classified as a psychophysiologic gastrointestinal reaction. "This diagnosis applies to specific types of gastro-intestinal disorders such as... ulcerative or mucous colitis... in which emotional factors play a causative role." [28] The pathogenesis of the disease in that case would be psychophysiological. Specifically, "This physiological dysfunction apparently involves parasympathetic stimulation of the lower bowel with production of the mucolytic enzyme lysozyme, which deprives the bowel of its protective coating of mucin. In certain personalities, this dysfunction may occur as a reaction to variety of stresses, but it is reputed to occur particularly in situations which demand independent accomplishment or arouse fear of not being able to do something." [29]

Such etiological ambiguity is not confined to this disease, nor to those specifically classified as psychophysiological reactions. In fact, with a bit of ingenuity, all diseases and functions of man can be seen to be infected with this same ambiguity. For example, heart disease is a disorder caused by emotional stress or life style as well as a disorder of the coronary arteries. All physiological processes and diseases occur within a social context so that they never possess a merely physical significance. This etiological ambiguity is an ever present possibility integral to the dual significance of man and man's world. He and his world are both physical and mental. The ambiguity is puzzling only for a dualist interpretation of the categorial structure of mind-body, which would construe mind and body to be competitors for causal efficacy in the field of being.[30] Properly understood, the ambiguity resides in the richness of the structure of mind-body which we have already indicated (see especially Chapter V C1). There is no causal competition; rather, the category of mental causality is a fuller realization of a physical cause.

---

[27] Lawrence C. Kolb, "The Psychoneuroses," *Cecil-Loeb Textbook of Medicine*, pp. 1717-1718.
[28] Committee on Nomenclature and Statistics of the American Psychiatric Association, *Mental Disorders*, American Psychiatric Association (Washington, 1968), p. 47.
[29] George Ulett and D. W. Goodrich, *A Synopsis of Contemporary Psychiatry* (St. Louis, 1969), p. 189.
[30] The ambiguity is not puzzling for a monist since he simply denies one of the dimensions requisite for the ambiguity.

Yet, in that certain diseases, such as ulcerative colitis, are designated "psychophysiological disorders," it is worthwhile specifically explicating the nomological structure involved. Psychiatry categorizes diseases as psychophysiological where it is particularly useful to designate a somatic state as having a psychical, not a somatic, origin. This can be expressed neutrally. "Frequently the nature of a body disorder can be appreciated only when psychological happenings, as well as physical disturbances, are investigated." [31] This "appreciation" could mean one of two things. First, it may be an appeal for a more complete cataloging of causes in order to find a gap filled by psychical agency. If there are such gaps, then a psychoanalytic account could provide missing links integral to a full (exhaustive) presentation of the etiological plexus. In the case of ulcerative colitis the issue would be whether, for example, an anxiety state, as an independent psychical cause, was necessary for the appearance of inflammatory infiltration and abscess formation at the base of the crypts of Lieberkühn, etc. But all references to the role of anxiety states can be translated into descriptions of the nervous system so that anxiety states can be integrated within the physical reality of the body. This would be something like "certain prolonged patterns of activity in the limbic cortex, etc., can cause a habitual discharge in the posterior hypothalamic area which when mediated to the lower colon (via the dorsal longitudinal fasciculus and then the lateral and ventral reticulospinal tracts and finally the pelvic nerve) causes an altered permeability of the capillaries of the colon with subsequent pathological changes pathognomonic of ulcerative colitis." In this case the description is categorially incomplete without the further understanding that the physiological activities of the cortex initiating the pattern of hypothalamic discharges are the physical substructure of a mental state. This, though, is the second and proper interpretation of this psychiatric category: certain physical dysfunctions of the body are fully understood only when their cause is grasped in terms of its psychical significance, and when this significance is articulated within a psychical nomological structure. That is, the assertion would be that one would not have fully understood the etiology of ulcerative colitis until he had apprehended its psychical causation and the role that such anxiety regularly plays in human psychodynamics. It is a matter of attending to the further meaning of certain causes, and is not a choice between alternative causes. The role of psychosomatic medicine is thus the further explanation of disease processes and their treatment by identi-

---

[31] A. P. Noyes and L. C. Kolb, *Modern Clinical Psychiatry* (Philadelphia, 1963), p. 381.

fying and attending to the psychological aspects of these diseases. It does not involve identifying or altering separate psychological processes integrated in an otherwise physical chain of causality, since all mundane causality is already physical.

This conclusion implies recognition, at least theoretically, of the possibility of determining physiopathological causes for all mental diseases and of finding specific psychopharmacological remedies for diseases now treatable only via psychotherapy. Of course a physiopathological basis for psychopathologies has been recognized with more or less clarity since the days of Hippocrates. But at present, a pattern of evidence is accumulating which indicates that biochemical defects underlie all mental diseases.[32] This is in consonance with our a priori analysis, but it introduces a theoretical difficulty, since there is also a tradition of psychoanalytic interpretations of the genesis of these diseases that omits reference to a physiological basis. This is not meant to imply, for example, that Freud was oblivious to the physical basis of psychical events. He surely was not: "We assume that mental life is the function of an apparatus to which we ascribe the characteristics of being extended in space and of being made up of several portions..."[33] But much Freudian analysis often seems to connote such obliviousness. For example, "in classic Freudian theory, schizophrenia is a regressive illness. Due to a failure to develop satisfactory object relationships during childhood, an adult developing schizophrenia shows withdrawal of libidinal cathexes from objects and a return to the earlier narcissistic level."[34] If we accept such psychoanalytic accounts as useful or even true, we must then explain what role they would play vis-a-vis a physiological account. We must also examine the status of psychoanalytic therapy, if there is a possibility of providing neuropharmacological therapy, etc., for all psychiatric disorders. We will sketch an outline of an answer, drawing upon our previous conclusions.

[32] This is most clearly seen in the evidence for a genetic basis for or diathesis towards mental illness – which strictly implies a biochemical basis. See for example Philip Solomon and Vernon D. Patch, ed., *Handbook of Psychiatry* (Los Altos, 1969), pp. 84-88. Also F. J. Kallmann, *Heredity in Health and Mental Disorder* (New York, 1953), and *Examining Goals in Genetics in Psychiatry* (New York, 1962) as well as Eliot Slater, "Genetical Factors in Neurosis," *British Journal of Psychology* 55, (1964), pp. 265-269.

There has also been much effort in delineating the basic biochemical aberrances involved. There is evidence, for example, to support the postulate that schizophrenia is pathophysiologically an autoimmunity disorder involving an antibody which has been termed "taraxein." See Robert Heath et alia, "Schizophrenia as an Immunological Disorder," *Archives of General Psychiatry* 16 (January, 1969), pp. 1-33.

[33] *The Complete Psychological Works of Sigmund Freud*, Vol. 23, *An Outline of Psycho-Analysis*, tr. James Strachey (London, 1964), p. 145.

[34] J. L. Reed, "Schizophrenic Thought Disorder: A Review and Hypothesis," *Comprehensive Psychiatry* 11, (September, 1970), pp. 403-432.

ONTOLOGICAL AND EMPIRICAL STRUCTURES   153

First, psychoanalytic or psychodynamic accounts can function, at least provisionally, to provide a more ready index of a person's pathological state. That is, it may prove easier to make diagnostic distinctions via psychiatric interviews rather than biochemical tests, at least for the foreseeable future. This would obtain especially in the case of neuroses where a psychological characterization of the illness presupposes far greater neurophysiological skills than are at present available.[35] Interestingly, Freud himself made this evaluation. "The future may teach us to exercise a direct influence, by means of particular chemical substances, on the amounts of energy and their distribution in the mental apparatus. It may be that there are other still undreamt-of possibilities of therapy. But for the moment we have nothing better at our disposal than the technique of psycho-analysis, and for that reason, in spite of its limitations, it should not be despised." [36]

Second, it is to be expected that psychological states be controllable via biochemical means; mind is embodied and thus dependent on matter (see Chapter IV B1). Nor is it to be unexpected that psychotherapy could also be understood as a reconditioning of neural circuits as an attempt to alter the patterns of response to the environment. But again, this does not spell the end of psychoanalysis. Nor is it merely granted a reprieve. Nor does its continuance lie in the fact that psychoanalysis can be redefined as (and *reduced* to!) a discipline concerned with the interconnection of generalized patterns of neurophysiological behavior and gross bodily behavior (and thus remain distinguishable from neurology, which is concerned with functions of the nervous system affected by focal or localized lesions, or general systemic disorders). Rather, psychoanalysis possesses an enduring importance because of the fact of man's psyche and the need to understand his body in terms of this psyche. Psychoanalysis (and psychology) provides an analysis of man in terms of his mental processes.

Psychology and psychoanalysis thus play a role that has both a scientific and a cultural importance. It is scientific in allowing one to apprehend a further nomological unification of physical events. Physiological processes take on a categorially novel significance; they become a new set of data. In so doing psychology (especially as psychoanalysis) serves the cultural goal of overcoming the otherwise alien otherness of the merely physiological. It tells the story of neurophysiological events in mental language, thus succeeding even when it fails to give an account

[35] At present little is known concerning the physiological basis of psychoneuroses, though constitutional factors are hypothesized as disposing towards them. See Lawrence C. Kolb, "The Psychoneuroses," p. 1710.
[36] *An Outline of Psycho-Analysis,* p. 182.

true to the facts. That is, it serves the purpose of giving a model or an empirical myth through which otherwise brute happenings are in some sense humanized. This is striking when the account employs elements of time-honored myths such as Freud's use of the Oedipus story. Of course the story becomes more powerful the more it is actually connected with "the facts." Thus the ideal remains an actual account of mental events and their own dynamics (e.g., laws of the association of ideas, constitution of inner time consciousness, repression of drives and formation of neurotic traits, etc.). Though these all have a physical basis they should also be understood as actual empirical psychical regularities, affinities and structures of appearance. That is, the study of the intrinsic regularities of psychic events is necessary for an encompassing account of man; onesidedness, as ever, must be avoided. As Freud himself pointed out, the reduction of psychic processes to their physiological substratum can "at the most afford an exact localization of the processes of consciousness and would give us no help towards understanding them." [37]

Thus, psychotherapy maintains its importance, even if neuropharmacology can supply us with the necessary and sufficient tools for controlling the brain, since psychoanalysis provides us with the understanding of this manipulation in terms of ourselves as subjects, not just objects. It is one thing to know the mechanism of controlling ourselves as objects and to have the tools to do this, and another thing to understand the significance of this in terms of empirical subjects. The first is the physical substrate and ground of the second. Efficient causality may be found in the first alone. But this does not mitigate the importance of the sciences in which certain efficient causes studied by the natural sciences are realized to be, in their full reality, also agents. Psychological or psychoanalytic explanation reveals the mental significance of otherwise alien processes, and at the very least provides an insight into the personal significance of our bodily processes. It is only when physiological and pathophysiological processes are understood in terms of their significance within the domain of mental life that they are comprehendable as a person's physiology or a person's disease. Otherwise, their significance is truncated and they are presented as impersonal occurrences. In short, it is through appreciating the physical processes of a body as fulfilled in the superstructured significance, mind, that it becomes possible to encounter a person or a patient rather than merely a specimen or preparation.

In this regard we must conclude by recognizing the unique function of psychophysiological (and physiological psychological) studies. They

[37] *Ibid.*, p. 144-145.

are the transilient empirical endeavor bridging two categorial levels by coordinating their events and the regularities of their events. Psychophysiology exhibits, as it were, the empirical features of the dialectical framework of mind and body. For example, psychophysiology attempts to show the physiology of fear without denying that there is indeed the psychical phenomenon fear. This is even the case if behavioral criteria for psychical states are employed, as long as one does not dogmatically deny the existence of the psychical states for which behavior acts as a sign. Thus via psychophysiological observations and theories, one can on the one hand recognize the psychical significance of neurophysiological processes, and on the other hand recognize the psychical significance of physiological processes. This is the singular importance of psychophysiology. As the author of the *Principles of Psychophysiology* puts it, "Perhaps the most basic question, implicit in the definition of the field, is: What is the relationship between the mind and the body? ... In the field of psychophysiology, however, we have virtually the only *experimental* attempt to find answers." [38] Psychophysiological studies [39] thus promise to specify the empirical parameters of embodiment and the criteria for the existence of minds (see Chapter V B1). For this the dialectical relation serves as the categorial necessary precondition.

2. Richer notions of human reality

But the appreciation of man's somatic life is not completed by apprehending physiological processes as reticula of drives, etc. Man's mental life is more than a complex of drives and impulses. Therefore we must add that psychology fulfills its role only insofar as it also clearly acknowledges the richer concepts of conscious life which are revealed, for example, by phenomenological psychology [40] and phenomenology of the cultural domains of human life: [41] cognitions and valuations, social life and action, etc. Psychiatric investigation and treatment cannot be just the reduction of human behavior to a plexus of subconscious forces. This has practical implications. For example, the role of values in attitudes

---

[38] Richard Sternbach (New York, 1966), p. 9.
[39] "Psychophysiology" is here used to designate the study of both the relation of the psychical to the physiological and the physiological to the psychical. In strict usage, psychophysiology refers to the study of the physiological correlates of psychical states, not to the study of the psychical correlates of physiological states. This latter is relegated to physiological psychology.
[40] See Edmund Husserl, *Husserliana*, Vol. 1, *Cartesianische Meditationen*, #11 and #16.
[41] *Ibid.*, #55 and #58, and *Husserliana*, Vol. 4, *Ideen II*, Beilagen XII and XIII.

towards suicide and dying cannot be assessed if the valuations are described only as affective expressions of subconscious complexes. One wants in addition to understand what values are held to be worth dying for, or which endow dying with meaning, to understand what circumstances and what ethical and valuational presuppositions make suicide appear as a reasonable choice.[42] This requires that one study patterns of conscious value-orientations and typologies of volition and action. This of course takes us beyond the individual to problems concerning sociology in that values and orientations towards values are involved at least at some juncture with the higher personal unities of the social world – family, society, state. Yet all along the importance of the physical moment remains, not only for theory, but also for practice. A political state can be poorly actualized not because of its political structure but because its citizens are plagued by intestinal parasites and are consequently lethargic. Socio-ethical life and physiology are not unrelated.

The importance of investigating man's whole significance comes particularly to the fore in the study of sociopathic personality disturbances. Their most basic and thus categorially impoverished significance is seen in a merely physiological understanding of the constitutional factors, conditioned patterns of neural activity, etc., that constitute the physical embodiment of sociopathic behavior. A second level of significance is seen in the traditional domain of psychoanalysis which recognizes the fuller significance of these physiological processes as impulse neuroses, etc. But the third level concerns sociopathic behavior as the failure to effect ethical conduct as well as a failure to realize personal existence within the higher unities of society and state.[43] This last level is the integrated comprehensive view of man which would embrace his reality as ethical and political man, and which presupposes his reality as psychological man (the latter of course also presupposes his reality as a physical reality). Only in terms of the fully personal level can man's life be understood as having a value for himself, and his fellow man. In its absence there are processes, inclinations, drives, and compulsions, but no values. The realm of values, in contrast, is the stance from which comprehensive judgments can be made concerning physical and psychical occurrences. It unites the other spheres of man's reality in terms of his

---

[42] Of interest in this regard is a number of articles in a collection, *The Meaning of Death*, ed. Herman Feifel (New York, 1965).

[43] For purposes of brevity the distinctions between ethical, social, and political reality will not be attended to. It is sufficient to note that each concerns values and can be studied by disciplines concerned with values, e.g., ethics, sociology, political science, and law.

reason, in terms of rational goals and purposes. Psychophysiological explanation consequently is more fully understood as ethicophysiological or sociophysiological explanation, since it encompasses man's reality as a being with values.

It is the ethicophysiological reality of man that most forcefully requires a richer model for psychotherapy than for medical therapy. Though anxiety has both pathophysiological and depth psychological substrata, it is often also a crisis involving values. Indeed, such a description indicates in an important sense the full reality of anxiety (and other such mental states), its higher truth so to speak. It is only because disease is a disvalue that enterprises such as medicine are initiated. And the singular mark of mental illness is that the derangement involves the very seat of the appreciation of values: conscious life (and its physiological substratum). Thus here more than elsewhere onesidedness makes a difference. Here treatment can most clearly be seen to be as "assistance" given to another person, not as merely the repair of a disordered physiology or psychology.[44] To be mentally distraught is to find one's freedom curtailed, to find special difficulties in social relations, to find independent activity in many ways encumbered – not by a predominantly physical handicap, or even by deep psychological forces, but by the strain of action under certain circumstances given certain values. One thus comes to see psychotherapy, as Szasz puts it, "as social action, not as healing. So conceived, psychoanalytic treatment is characterised by its aim – to increase the patient's knowledge of himself and others and hence his freedom of choice in the conduct of his life." [45] In appreciating the singularly human nature of such psychotherapy, though, one must not in his enthusiasm deny the pathophysiological and depth psychological substrata that underlie even the highest dimensions of human life. Caution, for example, is necessary in developing Szasz's position, which must otherwise tend towards a therapeutic dichotomy. The medical model fails to account for much of psychotherapy, not because it is wrong, but because it is not encompassing enough.[46] Models of therapy, medical and psy-

[44] Wolfgang Lederer, "Some Moral Dilemmas Encountered in Psychotherapy," *Psychiatry* 34 (February, 1971), pp. 75-85.
[45] Thomas S. Szasz, *The Ethics of Psychoanalysis: the Theory and Method of Autonomous Psychotherapy* (New York, 1965).
[46] What is implied here is that Szasz overreacts in labeling mental disease a myth. Mental disease is rather more than a disease. See also Szasz, *The Myth of Mental Illness* (New York, 1961). Recently D. A. Begelman has written an article reviewing the difficulties of the medical model in psychiatry and psychology. His delineation of the conflicts of models seems, at times, to appreciate the possibility of a unity in difference between the models. "Confirmation of the biogenic hypothesis does not necessarily preclude the truth of particular explanations embodying learning-theory concepts. In-

choanalytic, presuppose a way of thinking about the unity and diversity of man. In short, they underline the need for an encompassing and synthesizing view of man's reality, since medicine deals with him both as physical object and as free mind.

3. *The unity of human reality – the need for a categorial model*

In closing this chapter we can but hint at the present, pressing need for a notional account of the integrity of human reality. In many respects there is an incipient conceptual crisis: man is on the brink of experiences that will shock accepted patterns of ideas. For example, our scientific advances already are changing the biological or genetic essence of man. From this juncture evolution is no longer something beyond our control, but that which must progress in a particular direction either through our acts of omission or commission. We can allow our range of adaptation to be progressively restricted through the survival of persons totally dependent upon medical services, or we can strive to alter our genetic code in order to produce a "superior" species. But the notion of the compleat man is as ambiguous and as elusive as the notion of the superman. Yet the advance towards either is becoming at least a technical possibility. Man theoretically stands on the threshold of a brave new world in which the population explosion can be eliminated by artifical development of the population, physical defects and stupidity by modification of the genetic code, and untimely death by replacement of organs. We are about to attain the means for all of this while the goals remain undefined and the values involved obscure. Our technical achievements will in time seriously challenge our notions of human life. Not only will this require a richer notion of psychology and sociology, but it will charge these disciplines and the pure humanities with providing us an understanding of the destiny of man. We must learn not only to control ourselves and our environment, but we must as well learn to evaluate this control if we are not to become alienated from our own being.

It is no accident that the art of the twentieth century portrays a fractured or bifurcated reality in which understanding is separated from the forceful physical processes underlying action and life. The appreciation of the unity of psychical and physical reality has been shattered. This

deed, the truth of a biogenic hypothesis cannot conceivably preclude a learning-process etiology in its widest sense. To assume otherwise would be to confuse independent levels of discourse by supposing that biogenic and psychogenic theories are competitive or incompatible." "Misnaming, Metaphors, the Medical Model, and Some Muddles," *Psychiatry* 34 (February, 1971) p. 52.

problem is of such a nature that it cannot be solved piecemeal – indeed, the problem is that a piecemal solution is the source of our difficulties. Precisely what is needed is a unified and systematic view that allows interaction and separation of levels of inquiry so that science can be one without reducing its variety to a onesided monotone. Within such a unity of sciences, the categorial richness of mental and physical reality must be appreciated in its empirical concreteness. In short, we need a non-reductive view of the complexity of human reality. We must learn, for example, how to evaluate goodness in environment and potential in man – a recognition of the juncture of the value – free sciences and values. This point is made, for instance, by Kenneth Newell: "It appears time for us to spend some resources upon developing methods of classifying man, and of using our findings to develop a meaningful method of describing our ecology and relating it to the men who live there. It also appears time for us to stop saying that a "good" environment is one where the majority find it most comfortable. Maybe this research should be done almost as an emergency, before building our urban societies." [47]

This all goes to say that a more encompassing study of the relation of human physical and mental life (inclusive of the values it involves) is necessary if we are to develop what are often considered natural sciences: human physiology, ecology, and epidemiology. The last case is particularly illustrative of the need for empirically relating man's physical reality to his mental life, in particular human goals and values. Notions of human disease and health are essentially infected with values not only because they have roles within the psychodynamics of man (i.e., all are in some respect psychosomatic, see Chapter V D1) but also because "health" and "disease" are recognizable only with reference to man as a goal-oriented being. That is, if health is the ability to adapt successfully to one's environment, then definitions of "success" and "environment" are forthcoming, which recognize the complexity of human reality. Or put another way, the concept of "health" is complex and involves two major and distinct levels of significance. First, it refers to the physiological state of an organism, which can be expressed in terms of its ability to perform those functions that a) allow the organism to maintain itself ceteris paribus in the range of activity open to members of its species, and b) are effectively conducive towards maintenance of its species. Second, health can be understood in terms of the values especially associated with human life. Health in this sense is the ability to engage in

---

[47] Kenneth Newell, "Medicine and the 'New' Research," *The Bulletin of the Tulane University Medical Faculty* 26 (August, 1967), pp. 241-242.

those activities essential to realization of certain values. This second level allows further discrimination between health and disease. For example, conditions such as sickle cell anemia, which are successful adaptations to challenging environments, become recognized clearly as disease states, given man's reluctance to approve the metabolic doom of even a small percentage of his kind. Moreover, diseases which principally destroy higher mental functions appear as especially morbid, given the paramount value of intellectual abilities. In fact, only at this level do the notions of health and disease enable one to understand *why* disease is to be avoided. As a pure physical concept disease is but a negative direction in the economy of a species or the performance of an organism. Only in terms of a system of values can the worth of health and the disvalue of disease be assessed.

Such recognition of the ontological complexity of human existence cannot occur without one abandoning the attempt to reduce cultural sciences (the *Geisteswissenschaften* – both social and human) to value-free natural sciences, and thus acquit us of the responsibility of taking the cultural sciences seriously.[48] If values are but solemn expressions of emotion that effervesce around the periphery of objective truth, then it is clear why we are excused from taking them seriously.[49] But if value-infected dimensions of reality develop out of and complete those that are value-free, the conclusion is quite to the contrary. We must in that case explore what values are integral to a full human life, we must meticulously define mental health and disease. Then in terms of a notion of full mental health and development we can better judge what is, in general, a disease process, what is a bad environment, what should be the directions of our evolution.

In short, we are again returned to the notion of the dialectical unity of man – the organic interconnection of physical and mental reality. This model serves as the categorial justification for an amplified naturalistic attitude – for a notional understanding of how the range of scientific investigation can encompass two such diverse spheres as man's physical and psychological reality. It underwrites the empirical project of a broader, more unified investigation of man. In this fashion the categorial analysis of mind-body provides a general notional prototype for the relation of the

---

[48] This loss of contact with the realm of spirit, and the resultant crisis, is investigated by Husserl in *Die Krisis der europäischen Wissenschaften* (*Husserliana,* Vol. 6). Though his analysis does not proceed with a full consideration of the importance of natural science in the concrete realization of cultural sciences, it does offer an incisive perspective upon the problem.

[49] For example, A. J. Ayer, *Language Truth and Logic* (New York).

sciences concerned with man. This scope of application, moreover, further confirms the legitimacy and worth of such basic conceptual analysis.

E. CONCLUSION

> "The brain and the mind constitute a unity, and we may leave to the philosophers, who have separated them in thought, the task of putting them together again."
> Lord Brain, *Clinical Neurology*, p. 365

This treatise has attempted, as Lord Brain phrased it, to put mind and body back together again. It has been a presentation of the logic of two categories, an attempt to illuminate the conceptual architecture of a structure central to the character of being. The success of this endeavor is not, as Lord Brain's remarks suggest, a matter of indifference to science, much less to more general projects of understanding man's place in nature. On the contrary, as we have just noted (see Chapter V D3), an understanding of the logic of these categories is a prerequisite for grasping the general features of man's existence, in particular his ontological complexity. Science can remain aloof from such categorial questions only so long as it deals with a single categorial level and insofar as it tacitly presupposes the categorial structures with which it operates. As soon as science concerns itself with categorial structures that are blatantly complex, all description and correlation has a distressingly equivocal character. Categorial complexity involves a depth that opens the possibility of different ontological perspectives. Natural laws then manifest an ambiguity since they express the integration of ontologically multiform processes; they encompass different strata of significance. Mind-body and the question of the relation of physical and psychical descriptions and laws have been an example of this general quandary. It is such a quandary concerning the integration of different levels of significance that acts as a natural occasion for an exploration of categorial relations. One is moved to account for levels of nomological integration and to make sense of a complexity which must otherwise either be ignored or lead to ontological perplexities such as the dilemma of monism and dualism.

A categorial account sketches the logic of being, the grammar of existence, the rational structure of reality. It offers an insight into the basic ways in which existence is achieved. As we have said, it explains why reality is the way it is; a categorial account displays the rationale of being. It can then be termed a "grammar of being" in order to stress that categories are the signal coincidence of being and thought, that they are the key to the "usages" of existence, the rules for the presented pleomorphic modes of reality. By attending to categories and their relationships one focuses on unavoidable ontological structures. One examines how being must be in order to make sense and accomplish the act of existence. The attempt to order categories, to delineate their relation is thus, as we have shown, a logical endeavor that at once has ontological meaning. One is attempting to outline the notional relations integral to the presented significance of reality.

Ontology is the attempt to speak the language of being, to isolate those basic structures of reality that are also the basic structures of thought about reality. Difficulties in ontology are then difficulties we have in thinking about being, difficulties in finding in thought the structures we need in order to apprehend being. Such difficulties have been amply illustrated by the "mind-body problem" – a problem of comprehending the unity of distinct elements of reality that are often presented in disjoining contrast. What is needed are not generalizations based on further information concerning mind or body. What is sought is not a new fact but a new insight into thinking about whole classes of facts. The need is logical and the focus is upon being. From this need one can understand how ontology meets a rational goal and is not motivated towards the erection of dogmas. It is properly directed rather to discovering basic notional structures integral to comprehending being: categories. It seeks the categorial perspective from which being makes sense, the basic rational structure presupposed within certain types of facts. It is from the vantage point of this categorial perspective that we have attempted to secure insights into psychophysiological causality and laws, and more importantly into mind-body as such.

This endeavor is of immense importance as soon as one tries to make sense of something so complex as man. For example, the final significance of the descriptions within neurophysiology must be bracketed until the relation of the physical and the psychical is explained. Nor will it do to abjure such fundamental questioning in the name of a positivism that eschews basic ontology. The price of neglecting ontology is the fragmentation of knowledge or the reduction of the ontologically complex to ontologically simplest denominators. Categorial ontology proceeds from the given ontological complexity of reality and reaches towards the unities that comprehend diversities in all their richness. It is the project of saving for knowledge the richness of being by justifying the place in thought of categories found in experience.

To understand this project one must recognize the fallacy of reifying the categorial (see Chapter III A4) and then diagnose the role of this fallacy in the generation of what is usually termed the mind-body problem. To re-state the point, one cannot understand the relation between different strata of being, indeed one cannot even appreciate the purpose of such a quest for understanding, as long as one conceives of these strata as rival things competing for a position on the field of being. This has been the crux of the mind-body problem. The problem has usually been engendered by considering mind and body as distinct, though perhaps integrated substances. One then considers how they could be interrelated – causally, or as substance and accident, etc. The categorial revolution is, as it were, merely the recognition that reality is a whole that has often, in places, a complexity of significance – that it has various unities, some fuller and encompassing more depth than others. In this fashion mind is recognized as a further unity and richness of reality not ingredient in physical reality.

This, though, is recognized without positing *something* immaterial in the sense of ghostly and extramundane. Rather it is the recognition that the unity of certain physical objects is more than a merely physical unity and that they are thus to be understood as more than merely physical. This does not mean, of course, that they are not material. It means rather that they have a meaning and reality that transcends but does not leave behind a mere physical reality.

Everyday life attests to this. One confronts the living bodies of his fellows and recognizes persons such as he – beings in which physical reality is part of the life of a mind. In doing so one has no trouble in realizing that what one confronts is more than merely physical. Nor is one apt to think that what he confronts is not also physical; one can meet them in flesh and blood. Problems arise, though, when one tries to analyse this reality, to understand how it can fit together and thus how one can properly speak and think about it. This is where we came in. Analytical reasoning accustomed to simple distinctions and used to dealing with things and not categories is confronted by a paradox: the existing unity of two essentially distinct realities. The project has been to clarify the paradox by showing that the unity is a dialectical, categorial one. Since this project has dealt not with things but with the necessary categorial presupposition of things, this has been termed a transcendental ontology. Categorial analysis, transcendental ontology, is nothing more than the attempt to unravel the strata of being and to account for them in terms of the demands of reason.

It has already been mentioned that psychology correctly understood embraces and completes neurophysiology so that the otherwise onesided perspective of neurophysiology is seen within a fuller view. The full force of this fact must be appreciated for an understanding of the sciences and their futures. Man has too often not only dichotomized his being but also succumbed to a prevailing temptation to reduce himself to his physical parameters. To resist this temptation succesfully, man must amplify his perspective of the world and himself. He must try to pursue sciences such as biology, psychology, and sociology in what may appear to be practically useless fashions. He must try to understand the lifeworld of the animal, to delineate what is involved in his own psychical life – in legitimately psychological terms. Only when this is done can he escape from a tendency to dehumanize and devitalize himself as a result of trying to produce knowledge that is merely physical and thus readily appropriable by technology. For example, medical technology can efficiently cure man without helping understand what it is to be born, to live, to become ill, to be cured and even eventually to die. Twentieth Century man appears in the moment of his technological success to have forgotten why he wished to succeed in the first place. He has lost touch with the living threads of his own values; his command over himself and his world has become something that he no longer understands. He has become alienated from his very knowledge of himself, for while he has expanded his understanding of the physical reality that underlies his mental life, he has failed to acquire knowledge of the psychodynamics of the life of the technologist he has become. Man in physical control of himself is alienated from the expanses of physical reality, which he now surveys and controls.

But even if this existential urgency should pass, the epistemological gulf would persist and the intellectual need to complete a view that is otherwise singularly onesided would remain. One must study in full ethico-psychological terms what values can be realized, given that the human physical reality is as we know it is, or could be. To take a simple example, cancer of the breast could be totally eliminated if breast tissue were removed in routine procedures analogous to circumcision.[50] This is, of course, not without important implications for the possibilities of mother-child, as well as sexual relations. Though we understand fairly clearly what would be physically involved in such an enterprise, the ethico-psychological impact upon our body image idea of maternity and sexuality is far from clear. We have no established notion as to how essential or merely accidental such factors are. What would it mean for man to

[50] This is indeed considered by certain authorities on cancer – personal communication of Professor Albert Segaloff. See also A. S. Segaloff, "Progress in the Treatment of Breast Cancer." *The Treatment of Carcinoma of the Breast,* ed. Antony S. Jarrett (Palo Alto, Calif., 1967).

become a mammal without mammae? In short, we must try to determine what is involved in being alive, in thinking, in being a social being embodied in a physical world. Only then can we grasp the extent to which we are a function of our bodies and the extent to which our minds manifest the emergence of the freedom of spirit.

This book has indicated that the answers to such questions will be found when we engage in a categorial analysis of being. Here we have attempted such an analysis of the relation of mind and body. It can serve as a starting point for further and more ambitious investigations. In providing the ontology of a categorially complex being it supplies ways for thinking about the existing unity of ontological diversity. This difficult project of thinking dialectically must be pursued further. We must develop our understanding of the ways in which higher strata of significance are higher – of the way they offer new predicates of reality integrated in a new unity, completing, fulfilling, and making more self-explanatory the unity and predicates of a prior domain of being. We must categorize the parameters of the dialectic of the emergence of richer being and thus provide the criteria for a valid dialectical understanding of reality. With a general notion of what counts as a move towards categorial richness we can then diagnose the nature of ontological relations between categorially heterogeneous domains of being.

The claim has been that transcendental ontology affords further opportunities for understanding reality by providing new dimensions of explanation. In concerning itself with encompassing different levels of reality, and with understanding the rationale of their relationship, it opens up new breadth and depth to our understanding of Being. This returns us to the notions of *encompassment* and *completeness*. One is forced to pursue them since the goal of an all-inclusive physical explanation is not sufficient even if it were to be fully realized: it would not apprehend other dimensions of reality nor explain their relation. Explanation must encompass all domains of reality if it is not to be onesided. Indeed, the relations between domains must themselves be examined; one must come to understand the ordering of categories and justify them. Explanation finds a completion when its general structure is brought into conformity with certain directions of reason. It is in this regard that an account of the dialectic has been essential to our understanding the ordering of categories in terms of their encompassment. We have attempted to analyse the features of the progression of ontological richness in the step from the category of body to the category of mind and thus explain how this satisfies reason's nisus towards more self-explanatory categories. In examining the categorial relation of mind-body, something has thus been said about explanation in general, about the horizons of understanding, which when overlooked lead to substantial metaphysical quandaries – such as the mind-body "problem."

The moral is that such "problems" are not to be overcome by restricting the scope of explanation, by banishing ontology, but by coming to grips with the complexity of being. It is only in a radical richness of explanation that being is ever to be comprehended. The old problems of metaphysics can only be solved when ontology reapproaches the pleomorphic reticulum of being about us and attempts to grasp and understand the basic structures ingredient in it. The task of thought is to comprehend the significance of being and then to comprehend the significance of the basic categories of being's significance for thought. Thought's task is to discover rationality – to come to terms with itself – to think of its own thinking, to reason out its own rationality, especially that which concerns the content of its world. Ontology is therefore inextricably phenomenological in apprehending the significance of thought's world, and transcendental in apprehending the significance of its own categories of significance.

This examination of the relation of mind and body ends then with a general conclusion about ontology: it must be both phenomenological and transcendental. These leitmotifs of investigation initiated by Husserl and by Kant and Hegel have to a certain extent combined to offer a categorial viewpoint. But this is not an extra-mundane viewpoint. Philosophical thought is not an arcane endeavor, but springs as Aristotle knew from wonder about the world that surrounds us all. Here it was occasioned by the contrast of the realities of mind and body, the mystery of whose structure provoked an attempt to understand. Philosophy draws the broad blueprints of being in the hope that we shall find ourselves more intellectually at home in the reality in which we are enmeshed. Philosophy as such satisfies a need common to us all as thinkers. In helping us to think about being and to think about these thoughts, it does nothing immediately practical. But it does not, though, merely spin a yarn or step aside to play an esoteric game. Rather it engages in the serious task of trying to understand what it means to know the things we know. Phenomenological description and transcendental ontological analysis are but names for certain disciplined maneuvers of thought calculated to bring us to this goal.

Having reached this goal (at least in outline) with the realities of mind and body, we have secured a way of thinking. The consequence is a categorial scheme. *Cui bono?* It is good for anyone who wonders about the rules for thought fully confronting reality – indeed his own reality. As such it is "only" pure theory – a gratification of reason's need for explanation. But it also clarifies the basic theoretical points of departure for sciences more concerned with practical matters: neurophysiology and psychology – sciences that the praxis of medicine employs as it turns ever more towards remaking man. Finally, all of this is but an answer to the ever-present challenge to everyman – of knowing himself, of making some fundamental sense of his being, of having a categorial understanding.

INDEX

absolute mind, 25f., 95n., 96f., 106, 112f., 124
adequacy, 19f., 81f., 87f., 146f.
adequate experience, 19
appearance, 5, 8–10, 12–22, 24f., 28, 31–34, 39, 52, 66, 86f., 90, 92, 98, 129
Arieti, Silvano, 76n.
Aristotle, 18, 166
Begelman, D. A., 157n.
being, 6f., 12, 24, 26, 30, 94f., 97, 128, 137
Berkeley, George, 70, 77, 85
Brain, W. R. (Lord), 161
categorial account, 5–7, 60–62, 118, 125, 136n., 161–163
categorial relation, 12, 60f., 121, 127, 130, 145f., 148, 166
categorial turn, 82–84
category, 7f., 12f., 20, 22–28, 35f., 39, 66, 84–87, 91, 95–101, 105, 107f., 110, 112, 114, 120, 124, 134, 138
category, constitutive, 8f., 11f.
category of particularity, 8, 10–12
category of presence, 8, 10–12
completeness, 87f., 165
Cramer, Wolfgang, 48n.f.
Descartes, René, 64–68, 72, 78, 85
dialectic, 26, 89, 91, 97f., 100, 102. 104, 107f., 112, 114f., 118f., 122–124, 136, 142, 144, 146
dialectic, negative, 119f.
dialectic, positive, 119f.
dialectic, progressive, 92, 100, 104, 109, 118f., 123, 125, 133
dialectic, regressive, 92, 100, 104f., 108, 123, 125, 132f.
disease, 151, 157, 159f.
double aspect theory, 78–80
dualism, 64–66, 68–73, 75, 78, 82f., 116, 119, 132, 147, 150

efficient causality, 50, 52f., 145f., 154
ego, eidos, 15, 34n., 35–37
embodiment, 1, 42, 44, 46, 59, 107, 130, 132–137, 140, 143f.
empirical science, 18f., 38, 40n., 55f., 74f., 82, 88, 128–130, 147, 159f.
encompassment, 20, 22, 70, 81f., 84f., 87f., 96, 98, 115, 126, 146f., 165
Engels, Friedrich, 108n.
epiphenomenalist theory, 64
experience, 7, 15, 19–21, 25, 55, 93
explanation, 23, 91, 99, 103, 107, 113f., 147, 165
Farell, B. A., 74f.
Feifel, Herman, 156n.
Feigl, Herbert, 64, 72f., 79f.
final causality, 50, 52
Findlay, John N., viii, 27n.
Fodor, Jerry, 51
founding-founded relation (structure), 13, 28, 39, 46, 49, 55, 60–62, 66, 68–70, 81, 87, 89–91, 96, 98, 105, 108, 129, 148
Freedman, A. M., 149n.
Freud, Sigmund, 149, 152–154
Goodrich, D. W., 150n.
Gurwitsch, Aron, 34n.
Guyton, A. C., 58n.
Hartmann, Klaus, viii, ix, 24n., 91n.
Hartmann, Nicolai, 4n., 46n.
Hartshorne, Charles, 77, 140
health, 159f.
Heath, Robert, 152n.
Hegel, G. F. W., viii, 10n., 11, 15, 21n.f., 23f., 26n., 33n., 47n., 55, 66, 90n.–93n., 95n.f., 100, 102n.–104n., 106n., 113n., 116n.–118n., 119, 121n., 122, 123n., 127n.f., 130n., 131, 135n., 138n.f., 166
Hobbes, Thomas, 70
Hönigswald, Richard, 44n.

House, E. E., 59n., 112
Hume, David, 33n.f.
Husserl, Edmund, viii, 5n., 11n., 14n., 20n., 23, 32n.f., 37n.–40n., 45n., 46, 47n., 53n.f., 94, 136n., 141, 143, 155n., 160n., 166
identity in difference, 22, 34n., 49f., 61, 66, 73, 79, 81f., 121f., 145
identity pole, 30–33, 35, 49, 53, 61, 77, 109, 111, 122, 135
identity pole, object, 32–35, 52f., 59, 61
identity pole, subject, 32–36, 52, 59, 61, 135
intentionality, 32–34, 50–53
interaction, 48, 64, 66, 68f., 81–83, 118, 131, 145, 159
Kallman, F. J., 152n.
Kant, Immanuel, viii, 10f., 15, 20, 21n., 23f., 32, 33n., 46, 48, 85, 86n., 89n., 102n., 139n., 143n., 166
Kaplan, H. I., 149n.
Kolb, Lawrence C., 139n., 150n.f., 153n.
Körner, Stephan, 8n.
Lederer, Wolfgang, 157n.
Leibniz, Gottfried Wilhelm von, 64, 69, 85
Libet, B., 57n.
Lieb, Irwin C., vii
Malcolm, Norman, 146f., 148n.
Malebranche, Nicolas de, 64, 66, 68
moment, 13, 60, 90, 92, 101, 135f., 141, 143, 148
monism, 22, 64, 70–78, 82f., 116, 119f., 147
natural attitude, 37f.
nervous system, 41, 43–45, 56, 59, 71, 112, 134, 135n., 136, 140f., 151, 153
neurophysiology, 2f., 55, 59–61, 74, 81, 112, 130n., 146–148, 163f.
Newell, Kenneth, 159
Nikhilanada, Swami, 78n.
noumenon, 24
Noyes, A. P., 151n.
objective idealism, 115f.
objectivity, 24–26
ontology, 20–22, 28, 63, 65, 82f., 87f., 138, 162, 166
ontology, transcendental, 22f., 27, 89, 119, 126, 128f., 134, 163
panpsychism, 77, 117
Pansky, Ben, 59n., 112

parallelist theory, 64, 68
Patch, Vernon D., 152n.
Peirce, Charles S., 8n., 11n., 14n., 16, 25n.f., 92, 115–119
Penfield, William, 56–59
Pepper, Stephen C., 72n.
phenomenology, 7, 14f., 28, 40n., 75, 81, 86f., 114, 138, 155
phenomenology, eidetic, 15, 17, 20, 22, 81, 85, 125
Place, U. T., 71
Plato, 40n.
preestablished harmony, 69
presupposition, 90, 97
principle of concreteness, 100–104, 107
principle of full explanation, 103, 113f.
principle of ontological stability, 100, 103, 113
psychiatry, 130n., 151, 155, 157n.
psychoanalysis, 149, 151–154, 156f.
psychophysiology, 46n., 149–151, 154f., 157
psychosomatic medicine, 149–155
quid juris, 22, 123f.
Rademaker, Hans, 109n.
reduction, 59, 63f., 70–72, 76, 89f., 146, 153
Reed, J. L., 152n.
reification, 80f., 84, 127, 143
Schutz, Alfred, 42n., 44n.
Segaloff, Albert, 164n.
Sellars, Wilfred, 71, 73, 134
Sepp, E. K., 58n.
Sherrington, C. S., 59
Slater, Eliot, 152n.
Smart, J. J. C., 71, 72n., 75
Solomon, Philip, 152n.
Spinoza, Benedict de, 64, 65n., 66, 79f., 85
Sternbach, Richard, 155n.
sublation, 91, 113
Szasz, Thomas, 157
teleology, 51f., 111n., 135n., 146–148
tendency, 53, 61, 97–99, 110, 112
transcendental turn, 24, 81, 84, 125f.
transplantation, 136–139
Ulett, George, 150n.
volition, 1f., 57f., 142f.
Wittgenstein, Ludwig, 3, 27, 59, 75
Zaner, Richard M., viii, 42n., 90n.
Zetzel, Louis, 149